BESTSELLING BOOK SERIES

Technical Writing For Dummies®

TECHNICAL BRIEF

No matter what technical writing challenges you face, the following Technical Brief is the first step in the writing process. Feel free to make a copy of this Brief as you work on each new project. With just a little practice, it will help you to write with confidence and competence and propel your career!

About the Document

1. Type of document
2. Presentation context
3. Target date for completion

Reader Profile

4. Who are the readers?
   A. Are they technical, nontechnical, or a combination?
   B. Are they internal (to your company), external, or both?
   C. Do you have multi-level readers?
   D. If so, what percentage are there of each?
   E. What's their level of computer knowledge, if any?
5. What do the readers *need* to know about the topic?
   A. What's their level of the subject knowledge, if any?
   B. How do they process information?
   C. What jobs do they perform?
6. What's the readers' attitude toward the subject?

Key Issues

7. What are the key issues to convey? (**A** is the most important.)
   A. ______________________
   B. ______________________
   C. ______________________
   D. ______________________
   E. ______________________

(continued)

Technical Writing For Dummies®

Cheat Sheet

Project Team

8. Who's who on the project team?

____

____

____

____

____

9. What's the approval cycle? (**A** is the final reviewer.)

A. ______________________________

B. ______________________________

C. ______________________________

D. ______________________________

E. ______________________________

10. Will this document be paper, electronic, or a combination?

If electronic:

A. Modem speeds

B. Graphic cards

C. Monitors: resolution; color or black and white

D. Offbeat browser or old version of popular browser

E. Readability on a PC or Mac

F. Budget constraints

If paper:

A. Quantity

B. Professional printer or computer-generated

C. Binding

D. Finished page size

E. Budget constraints

Item 5308-9.
For more information about Hungry Minds, call 1-800-762-2974.

Praise for Technical Writing For Dummies

"Sometimes, the hardest thing about solving a problem is knowing a way to get started. If you are interested in technical writing, read this book. *Technical Writing For Dummies* does a great job with the fundamentals and provides numerous tips and templates you can start with, and continue to use as a technical writer."

— James R. Malanson, Director,
Workforce Planning, Training & Development,
Compaq Global Services, Stow, MA

"*Technical Writing For Dummies* is a must-have reference for both the aspiring and seasoned technical writer. Technical writing is one of the fastest growing professions in the Millennium. Sheryl Lindsell-Roberts has written a terrific book that addresses the strategic skills needed to succeed in technical writing regardless of the changing technology."

— Carol Szatkowski, CEO,
Clear Point Consultants, Inc, Beverly, MA

"*Technical Writing For Dummies* is an invaluable resource for technical folk who need to write effectively to communicate with both technical and, especially, nontechnical people."

— Norman Buck, President,
Coyote Technologies, Inc., Harvard, MA, and president of the International Computer Consultants Association

"As an adult educator, I was especially pleased to read this latest book by Sheryl Lindsell-Roberts, *Technical Writing For Dummies.* It offers clear and concise methods *and* it provides varied approaches for dealing with different learning styles. Readers will find this text to be very useable, presenting a natural flow of steps for compiling challenging and complex information — it almost makes me want to write a technical report just to try out the methods! Very nicely done!"

— Barbara A. Macaulay, Director, UMass Center for Professional Education, Westborough, MA

"With *Technical Writing For Dummies*, Sheryl Lindsell-Roberts demystifies an often intimidating subject with practical advice and wit. Even after 20 years as a writer, I was able to improve my style using her easy-to-understand tips and techniques."

— Thomas A. Ingrassia, Assistant Dean For Academic Affairs, Graduate School of Management, Clark University, Worcester, MA

"*Technical Writing For Dummies* practices what it preaches about using an approachable writing style. This book puts you on a good track for designing and writing documents your readers will really appreciate. It gives many useful tips (even for seasoned writers)."

— Greg Bartlett, Director of Documentation of The Mathworks, Inc. and President of Society for Documentation Professionals (SDP), Natick, MA

"Clear, concise, easy to understand documents are vital to the success of any company offering technical products. *Technical Writing For Dummies* lays out everything technical professionals need to produce documents their customers can understand, reducing expensive customer support calls and providing a valuable in-house document for employees. Sheryl Lindsell-Roberts does a remarkable job of balancing examples and practical advice as useful to those new to technical writing as it is to pros looking for ideas to apply to upcoming projects."

— Tonya Price, former OpenAir.com Director of On-line Marketing and President, Association of Internet Professionals-495 Massachusetts Chapter (`www.association.org`)

"It's refreshing and gratifying when someone who has toiled in the trenches can rise up and share her hard-earned wisdom in such an engaging way! Sheryl has "been there and done that" and we all can benefit from her experience and story-telling style. Her inclusion of the unique powers of electronic publishing is especially enlightening. I know that my writing will improve as a result; if I can only put the book down long enough to get back to the keyboard!!"

— Harry Pape, Director of the Information Services Group, Windmill International, Inc.

Technical Writing FOR DUMMIES®

by Sheryl Lindsell-Roberts

Hungry Minds™

HUNGRY MINDS, INC.

New York, NY ◆ Cleveland, OH ◆ Indianapolis, IN

Technical Writing For Dummies®

Published by
Hungry Minds, Inc.
909 Third Avenue
New York, NY 10022
www.hungryminds.com
www.dummies.com

Library of Congress Control Number: 00-110789

ISBN: 0-7645-5308-9

Printed in the United States of America

10 9 8 7 6 5 4 3 2

1O/QZ/QV/QS/IN

Distributed in the United States by Hungry Minds, Inc.

Distributed by CDG Books Canada Inc. for Canada; by Transworld Publishers Limited in the United Kingdom; by IDG Norge Books for Norway; by IDG Sweden Books for Sweden; by IDG Books Australia Publishing Corporation Pty. Ltd. for Australia and New Zealand; by TransQuest Publishers Pte Ltd. for Singapore, Malaysia, Thailand, Indonesia, and Hong Kong; by Gotop Information Inc. for Taiwan; by ICG Muse, Inc. for Japan; by Intersoft for South Africa; by Eyrolles for France; by International Thomson Publishing for Germany, Austria and Switzerland; by Distribuidora Cuspide for Argentina; by LR International for Brazil; by Galileo Libros for Chile; by Ediciones ZETA S.C.R. Ltda. for Peru; by WS Computer Publishing Corporation, Inc., for the Philippines; by Contemporanea de Ediciones for Venezuela; by Express Computer Distributors for the Caribbean and West Indies; by Micronesia Media Distributor, Inc. for Micronesia; by Chips Computadoras S.A. de C.V. for Mexico; by Editorial Norma de Panama S.A. for Panama; by American Bookshops for Finland.

For general information on Hungry Minds' products and services please contact our Customer Care Department within the U.S. at 800-762-2974, outside the U.S. at 317-572-3993 or fax 317-572-4002.

For sales inquiries and reseller information, including discounts, premium and bulk quantity sales, and foreign-language translations, please contact our Customer Care Department at 800-434-3422, fax 317-572-4002, or write to Hungry Minds, Inc., Attn: Customer Care Department, 10475 Crosspoint Boulevard, Indianapolis, IN 46256.

For information on licensing foreign or domestic rights, please contact our Sub-Rights Customer Care Department at 650-653-7098.

For information on using Hungry Minds' products and services in the classroom or for ordering examination copies, please contact our Educational Sales Department at 800-434-2086 or fax 317-572-4005.

Please contact our Public Relations Department at 212-884-5163 for press review copies or 212-884-5000 for author interviews and other publicity information or fax 212-884-5400.

About the Author

I'm fortunate to have a job that would be my hobby if it weren't my profession. I love to write. Between freelance business writing assignments and business writing seminars, I've written 18 books for the professional and humor markets.

Beyond that, I wear a lot of hats, just as you do. I'm a wife and the mother of two wonderful sons — Marc, an award-winning California architect, and Eric, a dedicated Maryland chiropractor. I live with my husband, Jon, in *Parnassus*, the incredible home in Marlborough, Massachusetts (outside of Boston), that Marc designed. However, if home is where you hang your hat, my hat is a cap of the New York Yankees.

When my life gets more complicated than it needs to be, my warm-weather nirvana is my 30-foot sailboat, *Worth th' Wait.* Jon and I are on board every weekend that the temperature rises above 60° — if the seas aren't too treacherous. (We've also been out there when they were too treacherous, but not by choice.) I don't bring my suitcase stuffed with clothes because there isn't room to put too much; I've learned to minimize. All I need is sunscreen, a few pairs of shorts, some T-shirts, and a good book. Columbus wanted to prove that the world was round, and Captain Kirk wanted "to boldly go where no man has gone before" — Jon and I merely want to leave our obligations and our harried lives on shore. Everyone needs a nirvana, even if it's a spot under a tree or the corner of a room.

When I'm not writing or sailing, I travel, paint (watercolors, not walls), garden, photograph nature, read, ski, eat strawberry cheesecake, and work out at the gym (after the cheesecake, I really need to). I try to live each day to the fullest!

Sheryl Lindsell-Roberts, M.A. and T.W.E.*
*Technical Writer Extraordinaire

Dedication

I dedicate this book to Jon — my truly wonderful and very patient husband. Everyone needs one special person who loves them for who they are and helps them to know that dreams do come true. To me, Jon is that special person. And our boat, *Worth th' Wait,* is the personification of our lives together. Jon is my anchor, and I'm the wind in his sails.

Author's Acknowledgments

I want to express my heartfelt thanks to my family (blood and extended) and to my dear friends. Without their love and support, I wouldn't be the person I am today — and I wouldn't be realizing my dreams.

I want to praise all the "Dummies" (and I say that with utmost respect) who made this book a reality. This is especially true of Jill Alexander the Great, my acquisitions editor, whom I thank for her steadfastness, sound advice, and confidence in me. I also appreciate the keen insights of Suzanne Snyder, project editor; Tina Sims, senior copy editor; and Karen Callahan, technical editor. And I extend thanks to two special subject matter experts who kept me honest in several of this book's chapters: Dr. Barry Kingsbury and Jennifer Lund.

Publisher's Acknowledgments

We're proud of this book; please send us your comments through our Hungry Minds Online Registration Form located at www.dummies.com.

Some of the people who helped bring this book to market include the following:

Acquisitions, Editorial, and Media Development

Project Editor: Suzanne Snyder

Acquisitions Editor: Jill Alexander

Senior Copy Editor: Tina Sims

Acquisitions Coordinator: Lauren Cundiff

Technical Editor: Karen Callahan

Senior Permissions Editor: Carmen Krikorian

Editorial Manager: Pam Mourouzis

Editorial Assistant: Carol Strickland

Cover Photos: VCP–FPG

Production

Project Coordinator: Nancee Reeves

Layout and Graphics: Amy Adrian, LeAndra Johnson, Jill Piscitelli, Jacque Schneider, Brian Torwelle, Jeremey Unger, Erin Zeltner

Proofreaders: Laura Albert, Andy Hollandbeck, Jennifer Mahern, Susan Moritz, Carl Pierce, Nancy Price, Marianne Santy

Indexer: Steve Rath

Special Help

Dr. Barry Kingsbury, Jennifer Lund

Hungry Minds Consumer Reference Group

Business: Kathleen A. Welton, Vice President and Publisher; Kevin Thornton, Acquisitions Manager

Cooking/Gardening: Jennifer Feldman, Associate Vice President and Publisher

Education/Reference: Diane Graves Steele, Vice President and Publisher

Lifestyles/Pets: Kathleen Nebenhaus, Vice President and Publisher; Tracy Boggier, Managing Editor

Travel: Michael Spring, Vice President and Publisher; Suzanne Jannetta, Editorial Director; Brice Gosnell, Publishing Director

Hungry Minds Consumer Editorial Services: Kathleen Nebenhaus, Vice President and Publisher; Kristin A. Cocks, Editorial Director; Cindy Kitchel, Editorial Director

Hungry Minds Consumer Production: Debbie Stailey, Production Director

Contents at a Glance

Cartoons at a Glance

By Rich Tennant

page 221

page 7

page 153

page 35

page 101

Cartoon Information:
Fax: 978-546-7747
E-Mail: richtennant@the5thwave.com
World Wide Web: www.the5thwave.com

Table of Contents

Appendix A: Punctuation Made Easy 253

Appendix B: Grammar's Not Grueling 265

Introduction

Man is still the most extraordinary computer of all.

—John F. Kennedy, 35th U.S. President

All technical people are called upon to write technical documents at some point in their careers. Therefore, your career depends on your ability to write and present your information clearly and distinctly. So this book is for you if . . .

- You're an engineer, scientist, computer programmer, or information technology specialist.
- You're involved in any other technical field.
- You're a professional technical writer.
- You're a college student who will enter a technical field.
- You shake and grunt like an unbalanced clothes dryer when you're asked to write a technical document.

Not all companies enjoy the benefit of having a technical writer on staff, and technical people struggle through the rigors of writing these documents on their own. Although this book won't reveal the formula for turning lead into gold and it won't unlock the secret of perpetual motion, it will serve as your easy-to-understand guide through the maze of writing technical documents — paper and electronic.

This book is also for professional technical writers. Professional technical writers come from all walks of life: teachers, musicians, journalists, scientists, and more. Technical writing services are sought in the United States, Europe, Asia, and Latin America. So whether you're a technical person who finds that technical writing is something you must do to advance your career or you're a professional technical writer looking to fine-tune your skills, you'll find this book to be invaluable to your professional growth and survival.

Skills, Not Frills

In Internet time, what's cutting edge today is history tomorrow. Therefore, this book isn't about software or applications. *This book is about strategy* — learning the skills you need to write energized technical documents that have the impact you want on your readers.

Even if technology didn't change so quickly, the most sophisticated software wouldn't generate a high-quality technical document; that's the responsibility of the writer. For example, if legendary writers such as Shakespeare, Chaucer, Poe, Twain, or Longfellow had computers, would they have been more successful? Of course not. They were all successful because they mastered the tools of their trade. *Technical Writing For Dummies* will help you master the tools of your trade and develop the skills you need to excel.

Preview of Coming Attractions

In each of my *For Dummies* books, I use interesting opening quotes to begin the chapters. Therefore, each chapter starts with a technology quote by one of the sages through the ages. When you read them, you may shake your head, wrinkle your brow, and just wonder what they were thinking. Here's a sneak preview of the five parts of this book:

Part I: What It Takes to Write Technical Documentation

Learn about the red-hot market of technical writing and how masterful technical writing will enhance your career — whether you're a technical person who's called upon to write documents or you're a professional technical writer.

This part introduces the Technical Brief — a key element in writing dynamite technical documents. The Technical Brief gets you jump started. Just as you wouldn't take a cross-country car trip without a map, you shouldn't start a technical writing project without filling out a Technical Brief. Once you use it, you'll wonder how you ever did without it. It will help you to get to know your readers (the term I use for paper documents) and users (the term I use for electronic documents), identify the key issues, and understand the executional considerations.

Part II: The Write Stuff

Most technical documents are a collaborative effort — even if it's just two people (a technical writer and reviewer). This part walks you through the steps of preparing an ironclad production schedule, brainstorming, outlining, drafting, editing, rewriting, and testing.

Also, in order to write technical documents that are valuable to your readers, your documents must have a strong visual impact and an appropriate tone. Whether your documents are paper or electronic, visual impact is what grabs the readers' attention, and a befitting tone gets the message across clearly. Last but not least, you want your documents to be remembered for the "write" reasons. Proofread! Proofread! Proofread!

Part III: Types of Technical Documents

In this part, you find tips for whiz-bang user manuals, abstracts, spec sheets, evaluation forms and questionnaires, executive summaries, and presentations that leave your audiences clamoring for more.

Part IV: Computers and More

No technical writing book would be complete without focusing on the power of the computer and the Internet. Although electronic documents follow many of the same basic guidelines as paper documents, they do have their own unique flavors. This part goes into detail about those unique flavors. It includes doing research online, creating sights and sounds, developing computer-based training (CBT) and Web-based training (WBT), and writing online help.

Part V: The Part of Tens

The Part of Tens is a *For Dummies* classic. Here you find a potpourri of tips and tidbits in a variety of specific areas such as publishing in a technical journal, filing a patent, and writing a grant proposal. It also highlights ten ways to make your document shout "Read me."

The appendixes round out *Technical Writing For Dummies* with a glossary of technical terms you can bandy about at cocktail parties, punctuation and grammar guidelines so you can write to your readers in your voice, information on abbreviations, and tables of metric equivalents.

Getting the most from this book

I strongly suggest that you read Chapters 2 through 7 in sequential order because good writing is a process of getting started, creating an outline, writing the draft, designing for visual impact, honing the tone, and proofreading. When you work collaboratively, you may repeat parts of this process as needed.

The remainder of this book builds on this process for print and electronic technical documents. Feel free to jump to whatever topic interests you or applies to the writing challenges you face.

Icons, Icons Everywhere

To help you find the important stuff easily, I scatter icons throughout this book — somewhat like road signs. Each of the following icons pinpoints something vital to your technical writing existence:

The Sheryl Says icon helps you benefit from my experiences — the blissful, the painful, and everything in between.

The Tip icon gives you nifty tips to take on the road to effective technical writing. These may be time savers, frustration savers, lifesavers, or just about anything else.

The Remember icon represents little tidbits to — what else? — remember.

The Technical Brief icon reminds you to refer to the Technical Brief during different stages of your technical writing project. I detail the Technical Brief in Chapter 2 and present it on the Cheat Sheet in the front of this book so you can find it easily.

The Caution icon calls attention to a pitfall you should avoid. If you don't heed the caution, it won't put an end to civilization as we know it. However, forewarned is forearmed.

The Success Story icon plays off the adage "Nothing succeeds like success." You may find it helpful to hear other people's success stories.

Author's Note about Genders

When I started writing *For Dummies* books, I searched for an elegant pronoun that would cover both genders. I wasn't able to find one. Rather than get into the clumsy he/she or him/her scenario, I opted to be an equal-opportunity writer. I tossed a coin, and here's how it landed. (If this approach offends anyone, I sincerely apologize.)

- The male gender appears in the even chapters.
- The female gender appears in the odd chapters.

Making This Book Your Personal Reference Source

Following are a few ways to personalize this book so it truly becomes your reference source:

- Write your own notes in the margins; there's plenty of room.
- Highlight the stuff that's meaningful to you with a colored highlight marker.
- Get stick-on notes and tape flags to tag the hot pages.

And don't forget to put your name in big letters in some obvious spot. Books such as these have a tendency to find new homes.

Part I

What It Takes to Write Technical Documentation

In this part . . .

Just as there are two sides to every story, there are two sides to every technical document — the writer's side (you) and the reader's side (your audience). People who write for entertainers learn the writer-audience concept early on. The entertainers who get standing ovations, rather than catcalls, are the ones with writers who take the time to get to know the audience.

This part links the people who write technical documents (and that's just about everyone who's in a technical field) with those who read technical documents in paper or electronic forms, or a combination of the two.

This part also features the Technical Brief, which is key to understanding your readers. The Technical Brief helps to ensure that *your* technical documents get standing (or sitting) "O's."

Chapter 1

Accelerating Your Career the "Write" Way

In This Chapter

- Discovering who writes technical documents
- Understanding how technical documents differ from business documents
- Developing a strategic approach to technical writing
- Writing technical documents that have impact

640K [of memory] ought to be enough for anybody.

—Bill Gates, 1981

Whether you realize it or not, technical documents are part of our everyday lives — both personal and business. When you buy a new camera, it comes with instructions on everything from changing the batteries to getting rid of red eye. When you get a prescription from a pharmacy, it comes with a leaflet on how often to take the medication and what the side effects may be. When you hire an architect to design your home or office, the architect presents you with drawings of the layout. Technical documents are written for all of us, not just for computer geeks who assemble rockets or plasma generators. And it's not just the computer geeks who write technical documents — all technical people do at some point in their careers.

Technical writing means different things to different people. It covers the fields of electronics, aircraft, computer manufacturing and software development, chemical and pharmaceuticals, technical publications, health, and much more. It spans the public and private sectors as well as academic institutions.

Technical Writers Spring from All Walks of Life

People who write technical documents come from all walks of life — and most aren't technical writers per se. Here are some actual situations of people who were called upon to write technical documents in the course of their professions:

- **Computer programmer:** Pat graduated with a degree in computer science and was hired as a software developer for a company in the fast track. Several months later, the company felt a financial pinch and laid off the technical writers. Pat had a big deliverable due in a few months, and her supervisor told her that she had to write a user manual. Sophomore English (which Pat struggled through and loathed) didn't prepare her for this type of assignment. After all, Shakespeare wasn't a technical sort of guy. Poor Pat had to muddle through writing the user manual and got gray hair prematurely.
- **Manufacturing specialist:** Bill worked for a manufacturing company for many years and developed a piece of equipment that was expected to revolutionize the industry. The equipment made its debut in Germany at the industry's largest conference. Bill's supervisor asked him to deliver a paper (the industry term for a making technical presentation) at the conference. The audience would consist of more than 200 top industry professionals. Not only did Bill fear the podium more than the dentist's drill, he didn't know to prepare or deliver a technical paper — especially in a foreign country for an audience of this caliber.
- **Mad scientist:** While working at a pharmaceutical company, Nathaniel had a major breakthrough on a treatment that promised to prevent baldness. The company president asked him to write an article for a major medical journal. Although Nathaniel was flattered by the president's request, he didn't know the first thing about writing or submitting a technical article.
- **Sales representative:** Lynette was a sales representative for a worldwide computer distributor. She'd often be away from home for weeks at a time. After 15 years as a road warrior, Lynette suffered from burnout. (She used to leave her picture on the fireplace mantle so that her family wouldn't forget her.) Lynette had been reading about the burgeoning field of tech writing. She called a local college, got all the literature, and decided to pursue a master's degree in technical writing.

Although I changed the names to protect the innocent (Lynette, for example, hasn't turned in her resignation yet), scenarios such as these are typical. Technical people who aren't trained writers are constantly asked to write technical documents. Their education and work experience rarely prepare them for this type of challenge. *Technical Writing For Dummies* will!

The humble beginnings of tech writing

Technical writing as we know it today took root in World War II when the U.S. military persuaded "those who served" to write manuals to aid the war effort. The military needed to teach soldiers about weapons, transport vehicles, and other hardware. These "technical writers" had little or no training. They just sat down at their manual typewriters and banged out whatever made sense to them. I don't know whether it made sense to the poor soldiers trying to decipher their writing. However, we did win the war.

Technical Writing Differs from Business Writing

Many people ask the difference between business writing and technical writing. The difference is analogous to apples and oranges. For example, at the very core (pardon the pun), apples and oranges are fruits. And at the very core, documents are words and graphics. Beyond the core, business and technical documents are different species.

Documents of the business kind

Letters are the crux of business documents. When you factor in e-mail messages, that accounts for as much as 90 percent of all business correspondence. Every businessperson writes business documents — letters, memos, e-mail messages, proposals, reports, and more. One major difference between business and technical documents is that business documents are generally written by one person, often for a single reader or small, select group of readers. Following are some commonly written business documents:

- Agendas
- E-mail messages
- Letters
- Meeting minutes
- Memos
- Presentations
- Proposals
- Reports

For a super book on writing business documents, check out my hot seller *Business Writing for Dummies* (Hungry Minds, Inc.). This book is the outgrowth of my very successful business writing seminar, "Energize Your Business Writing," and it walks you through the Six Steps that are key to having the impact you want to have on your readers. It also ties those steps into all sorts of business writing, including those just mentioned. The book even includes a major part on using e-mail effectively and cutting information overload.

For expert letters, check out my other hot seller, *Writing Business Letters For Dummies* (Hungry Minds, Inc.). It's chock-full of ready-to-use business letters and e-mail messages for all occasions. I just got a letter from someone who said, "My version of your book has so many dog-eared pages and highlights that I can barely read it! This book is never more than two feet from my computer monitor. I have saved hundreds of hours, added to my letters more flavors than Baskin-Robbins, and learned the essentials of great letter writing. Get it NOW!"

Documents of the technical kind

People in specialized fields write documents that relate to technical or complex subjects. Unlike business documents that are generally written by one person, technical documents are often a collaborative effort between a writer, subject matter expert (SME), editor, and others. (Check out Chapter 3 to learn more about writing as part of a collaborative effort.) Technical documents are generally intended for lots of readers. Following are some commonly written technical documents — paper and electronic. You find chapters about the specifics of writing each of these documents later in this book.

- Abstracts
- Articles for publication
- Computer-based training (CBT)
- Evaluation forms
- Executive summaries
- Functional and detail specifications
- Online help
- Presentations
- Questionnaires
- Reports

- Training material (paper or electronic)
- User manuals
- Web-based training (WBT)

Documents such as reports, proposals, and presentations can be business or technical. *On the business side,* an advertising agency may prepare a colorful presentation to dazzle a client with a creative ad campaign. *On the technical side,* an engineer may prepare a conservative presentation to persuade management that her project needs additional funding. Check out Chapter 12 for some handy-dandy tips on preparing and delivering a dynamic technical presentation.

Print or Electronic Media — That Is the Question

Technical writing covers both print and electronic media and you must understand which (or a combination of both) is suitable for your reader. Following is a sampling of a few types of print and electronic media that fall under the broad category of technical writing:

- **Print**
  - User/reference manuals for hardware or software
  - Equipment specifications for people who assemble, operate, or repair machinery
  - Scientific articles, reports, and white papers
  - Papers to be delivered at seminars or conferences
- **Electronic**
  - Web-based documents
  - Computer-based documents
  - Online documentation (with Help included)

In the early 1990s companies delivered print versions of user manuals, parts catalogs, specifications, and the like. Now these same documents may be delivered in print or electronic form. The key — as I stress over and over in this book — is to understand the needs and environment of your readers. For example, many environments, such as semiconductor clean rooms or electronically run manufacturing plants and stockrooms, are completely paperless. For a great book on the latest trends, check out *Customer Service on the Internet,* by James Sterne (Wiley Computer Publishing).

Assigning Responsibility for Technical Documents

The responsibility for writing technical documents depends on a company's structure and resources. Following are three ways that companies typically generate technical documents:

1. **Technical gurus (engineers, software developers, and others) write their own documents.** Some of these people may have taken writing courses, but most have no training in writing a cohesive document. These "technical writers" often overlook steps to share with their readers because these steps are obvious to them. And they probably haven't identified the needs of their readers. They write what's important to them.
2. **These same gurus may draft documents and then turn the drafts over to technical writers to edit, format, and polish.** Unless the technical writer has an opportunity to learn the subject matter intimately, many of the steps that may have been overlooked by the guru aren't identified by the writer or editor. This process does, however, produce a document that may be more pleasing to the eye — for what that's worth.
3. **A technical writer is called in from the onset of a project.** The writer works with the developer who's the subject matter expert (SME). They work as a collaborative team, each adding their expertise to the project. This approach is generally the best of all possible worlds.

Strategy, Not Software

Anyone who writes technical documents must understand how critical it is to take a strategic approach. For example, if you design a custom home, do you first call someone to wield a hammer? Of course not. A hammer is merely a tool. To design a custom home, you call an architect — a trained professional who designs layout; renders plans for the plumbing, electrical, and heating systems; and provides the structure. Then you call someone who knows hammers.

The same holds true in technical writing. Effective technical documents require an information architect — *a technical writer.* Whether this person is a professional technical writer or an engineer or software developer who writes technical documents, she must plan, design, and provide logical structure. Anyone can learn to use the software to create the document. Much like the hammer, software is merely a tool. The key to writing a great document is *strategy,* not software.

Someone once told me that she wouldn't make a good technical writer because she can't even use jumper cables to rev up an ailing car battery. Remember that technical writing isn't about jumper cables or about understanding every aspect of the technical and scientific communities. And it isn't about knowing every nuance of the latest software application. Very few people have that broad a knowledge base. Technical writing is about using *strategy and resources* to write clear, accurate, and logical documents. If you apply a logical strategy and avail yourself of resources, you can write just about anything — from turning on your computer to assembling a jet airplane.

What You Need to Succeed

Following is a snapshot of what it takes to write clear and understandable technical documents:

- **Show respect for your reader.** You must always write with respect for your reader. Write with a positive attitude, not with arrogance. I've heard technical writers speak of their readers arrogantly with remarks such as, "We don't write these documents for idiots. If they're that stupid, they shouldn't be using this product." (Well, excuuuuse me.) Even the brightest humanoid may experience confusion when presented with something entirely new.

- **Pay keen attention to details.** You show your keen eye for detail in the way you think about what you write. I recently saw a resume from someone who was applying for a position as a technical writer. Check out these bulleted items I extracted from that person's resume:
    - **"Contributing writer to weekly city newspaper."** (Which newspaper? What does he write about? The newspaper is weekly, but is his contribution weekly? Bimonthly? Monthly?)
    - **"Generated technical reports."** (What does he mean by "generated"? Did he write the report or merely click the Print button and spew them out?)

    This writer wasn't able to identify the details that a potential employer wants answered. Therefore, it was obvious that he couldn't identify the details that his audience needs. *You must be able to anticipate the questions your readers will ask, and you must answer them.*

- **Know your readers and their requirements.** Unlike a letter that you may write for a specific reader, you often write a technical document for a diverse group of readers. You must understand their needs in order to determine how you write the document and whether print or electronic media is appropriate. Check out Chapter 2 for information on understanding your reader by using the invaluable Technical Brief.

- **Collaborative efforts.** To tweak John Donne's famous quote: *No tech writer is an island.* Even if you're the only tech writer on your project, you'll work with people inside or outside the organization: SMEs, editors, publication/electronic specialists, and end users. Chapter 3 talks about what it takes to work collaboratively.
- **Demonstrate the ability to leg-o your ego.** Last but not least, you must be able to leave your ego at the door. Your finished document will often be very different from your original draft. Everyone who reviews the document feels compelled to pick up a pen and mark it up. It's all part of the process — the need to make a contribution. So, be prepared to have your work edited, re-edited, and perhaps ripped to shreds. (There's no chapter on the latter. Just smile a lot and mumble to yourself.)

Seeing Is Believing

As mentioned earlier in this chapter, we come in contact with technical documents in our daily lives. Some documents are well written; others leave us scratching our heads and wrinkling our brows. What's the difference?

Following are *Before* and *After* cases that give you a chance to look at two different approaches for writing and presenting technical documents. I took each *Before* case from an actual document where the writer buried the key information. Each *After* case shows an improved version where the key information jumps out. (Check out Chapter 5 to discover how to visually prepare technical documents that call the reader's attention to the key issues at a glance.)

Case 1

The United States Air Force tabulated the weight of 80 officers. Following are key questions readers may have:

- What's the median weight?
- What's the highest weight?
- What's the lowest weight?

Before

The following chart is from the *ESD Process Improvement Guide* prepared by the Electronic System Division, Air Force Systems Command, Hanscom Air Force Base, Massachusetts. It displays the weights of the 80 officers in column format. The reader must search through all 80 numbers to answer the questions.

Weights of 80 Officers

206	**180**	**139**	**163**	**159**
155	**180**	**165**	**149**	**127**
159	**171**	**141**	**190**	**159**
153	**181**	**180**	**137**	**161**
115	**156**	**173**	**165**	**191**
159	**110**	**179**	**145**	**144**
150	**206**	**166**	**188**	**165**
127	**130**	**172**	**180**	**147**
145	**150**	**156**	**171**	**189**
190	**200**	**208**	**169**	**139**
130	**128**	**155**	**185**	**166**
165	**187**	**159**	**178**	**169**
147	**150**	**201**	**128**	**170**
189	**163**	**150**	**158**	**180**
139	**149**	**185**	**129**	**169**
175	**189**	**150**	**201**	**175**

After

The distribution of the data in the form of a histogram shows the answers to the readers' questions at a glance.

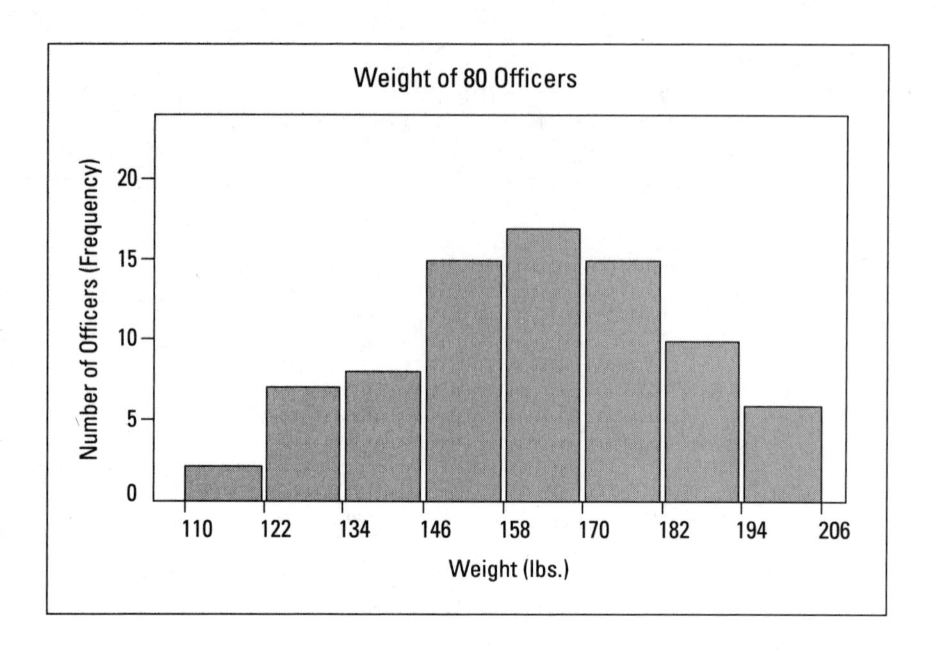

Case 2

The U.S. Department of Transportation wrote a "Transit Security Procedures Guide." Key issues in the following paragraph would be: What action must authorities take to prevent similar incidents? The Before paragraph buries this information; the After paragraph creates bullets to call attention to it.

Before

Training must include the sensitive handling of complaints and reports of offenses. Although these offenses are not actual assaults, there are victims. Authorized personnel must respond in a manner that leads the victim to believe that a similar incident will never happen again. This includes taking immediate action to apprehend the offender, taking information from the victim that may lead to arrest and prosecution, and notifying the police.

After

Training must include the sensitive handling of complaints and reports of offenses. Although these offenses are not actual assaults, there are victims. Authorized personnel must respond in a manner that leads the victim to believe that a similar incident will never happen again. This includes

- Taking immediate action to apprehend the offender
- Taking information from the victim that may lead to arrest and prosecution
- Notifying the police

Case 3

Readers of this technical report want one question answered. What are the findings? In the Before case, the reader must read through an entire paragraph to find the answer. In the After case, the answer is part of the headline and identifiable at a glance.

Before

Conclusion

The engineering analysis and observation did provide significant information. Review of the data indicates that a 10- and 12-mil printing process can produce consistent and acceptable result. The solderpaste heights were consistent and had a low standard deviation. The 8-mil spacing solderpaste heights were much lower with a larger standard deviation. Solder shorts and opens were much lower for 1- and 12-mil spacing as compared to the 8 mil. From a statistical point of view, the analysis indicates that no statistically valid conclusion may be drawn, due to the error in the model. Therefore, we must perform additional tests.

After

Conclusion: Due to an Error in Model, We Must Perform More Tests

The engineering analysis and observation did provide significant information. Review of the data indicates that a 10- and 12-mil printing process can produce consistent and acceptable result. The solderpaste heights were consistent and had a low standard deviation. The 8-mil spacing solderpaste heights were much lower with a larger standard deviation. Solder shorts and opens were much lower for 1- and 12-mil spacing as compared to the 8 mil. From a statistical point of view, the analysis indicates that no statistically valid conclusion may be drawn, due to the error in the model. Therefore, additional testing must be performed.

Saga of a mad tech writer

Following is the epic tale of how I got started writing technical documents. I can laugh at the scenario now, but it wasn't so funny at the time.

My technical rite of passage

Once upon a time I experienced the technical rite of passage — I bought my first computer. This was back in the days when the information superhighway was just a dirt road. The dealer assured me that the computer was "user friendly." I since learned that the expression "user friendly" has outlived its meaning. I set up my computer with the help of my then teenage sons and eagerly loaded the program disk according to the directions in the user manual. Wow, I thought, I cleared the first hurdle. This will be a piece of cake!

Let the frustration begin

With great confidence I read through the user manual and followed the step-by-step instructions. I keyboarded (the computer-age word for typing) some text and wanted to practice resetting the tabs. I checked the index and found the pages with instructions on setting tabs. I read through 2½ pages of instructions trying to figure out what the writer was telling me.

No matter what I did and how many times I did it, I couldn't reset the tabs. The computer just kept beeping. I was starting to feel inadequate. My hands got clammy. My heart raced. I started to twitch.

I thought the computer must be telling me: "Try again dummy." So, I re-read the instructions carefully and once again did exactly what I read. The beeping just wouldn't stop. I was determined not to call my sons for help. "Perhaps I need a first grade primer," I thought. I was so tense that I felt like a potato chip about to snap. I decided to call the dealer from whom I bought my computer. In just a few seconds he explained how to reset the tabs. For this I needed a 2½-page explanation of gobbledygook?

I could go on endlessly with the frustrations of trying to follow the user manual, but that would be a book by itself.

Words my mother taught me

Somewhere in my memory bank I recalled my mother telling me, "Don't complain about something unless you can do it better." That's exactly what I set out to do — write technical documents, and write them better.

Building a better mousetrap

I spent years honing my technical writing skills. I'm not exactly as old as Methuselah, but back then there weren't any books on the subject, and there weren't schools offering courses. I learned by trial and error, and my early clients were my guinea pigs. (I kept getting assignments and referrals, so I guess I did a pretty good job.)

With a lot of determination, I learned to write technical documents that readers rave about and I received several awards. I went on to teach technical writing workshops, and now I'm writing this book.

I did it; you can too!

Chapter 2

The Person to Whom You're Speaking

In This Chapter

- Understanding the fine points of a Technical Brief
- Filling out a Technical Brief

> *So we went to Atari and said, "Hey, we've got this amazing thing, even built with some of your parts, and what do you think about funding us? Or we'll give it to you. We just want to do it. Pay our salary, we'll come to work for you." And they said, "No." So then we went to Hewlett-Packard, and they said, "Hey, we don't need you. You haven't gotten through college yet."*
>
> —Steve Jobs, Apple Computer, Inc. (1976)

The key to writing an effective technical document is to have an in-depth understanding of your readers. You must understand who your readers are, the level of detail they need, how they process information, and how they'll use the document. If you don't gather that information, your document will be as ineffective as the foreign language directions that came with your VCR. The Technical Brief (discussed later in this chapter) is your key to unlocking this knowledge.

How to Feed a Martian

This is an exercise I use when I present technical writing workshops. At first, participants think I'm crazy (and you may too), but it does prove a very critical point. Please stay with me here.

Imagine this scenario: It's the year 2500, and business is conducted intergalactically. You have a hot deal pending, and it's hinging on the arrival of a business associate named Zeb from the planet Zeblonia. This is Zeb's first trip to the planet Earth. If you can razzle-dazzle him and get him to sign the deal, you'll get the big promotion you've been coveting.

You plan to be at the space pad where Zeb's spaceship will land. And you plan to welcome him personally and bring him directly to your office for lunch, followed by the all-important meeting. However, Zeb's ship is delayed because of heavy intergalactic traffic, and you can't wait at the landing pad. A crisis demands your immediate attention. You cancel the luncheon and reschedule the meeting for later in the day. You arrange for a driver to drop Zeb off at your house. This will give Zeb time to get a bite to eat and freshen up before he meets with you and the other earthlings.

You realize that Zeb will probably be starved after his long and arduous journey; after all he missed lunch. You didn't have time to shop (you weren't expecting him at your home), but you do want to make Zeb feel at home. The fixings for a peanut butter and jelly sandwich are all you have in the cupboard. Although it's not a gourmet meal, you think it will hold him over until later. You know that Zeb speaks a little English, so you leave him directions to make a peanut butter sandwich.

Your assignment, should you choose to accept it, is to write detailed instructions for Zeb on how to make a peanut butter and jelly sandwich. *I hope you'll try this exercise because there's a method to my madness.*

How to make a peanut butter and jelly sandwich.

____

____

____

____

____

____

Will Zeb go hungry?

What did you learn from this exercise? Perhaps nothing, but I hope that isn't the case. Check out some questions you'd need to think about in order to write clear instructions for Zeb to have a shot at understanding you:

- **What's Zeb's level of understanding?**
  - Although he speaks English, would he necessarily understand your terminology?
  - Would Zeb know a jar of peanut butter from a jar of jelly?
  - Would he even know what a jar is?

- **Did you give Zeb all the information (and steps) he needs?**
  - Would he understand how to remove the lid from the jar?
  - Would he even know he should remove the lid?
  - Is your peanut butter from a health food store? If so, did you tell Zeb to mix the oil with the gooey stuff?
- **Did you present the information for quick understanding?**
  - Will Zeb understand your terminology?
  - Would visuals help? (Perhaps you could draw a picture of the peanut butter jar.)
  - Would a combination of words and graphics be appropriate?
- **Did you achieve your purpose?**
  - Was Zeb able to make a peanut butter and jelly sandwich?
  - Was he smiling?

Filling Zeb's empty stomach

Most participants start out thinking that this exercise is quite easy. They begin with basic directions much like these. (It's important to number steps in a process, and not use bullets. More of that in Chapter 5.)

1. **Open the jar of peanut butter.**
2. **Open the jar of jelly.**
3. **Smear the peanut butter and jelly on the bread.**

Okay, these instructions may be simple for the average Earthling, but would Zeb know a jelly jar from a loaf of bread? In the scenario, I said that Zeb speaks English, but do you understand everything you read just because you speak English? For example, would you be able to distinguish *montmorillonite* from another mineral along side it? Do you even know what the word means? Just because you speak English doesn't mean you understand everything you see in writing. You understand only what you've been exposed to.

Of course, you won't be writing technical documents for aliens from another planet, but much of what you write may be alien to readers on this planet. It's vital that you understand your reader.

Getting Jump-Started with the Technical Brief

Before you write a technical document, you must understand how and why people read the documents. They don't read technical documents as they read *War and Peace.* They don't put their feet up on the table in front of a roaring fire and bury themselves in every pearly word. And they don't read technical documents for pleasure.

People generally use technical documents as references, often to figure out what they did wrong. In that case, they're frustrated and don't want to waste their time poring over gobbledygook. (Chapter 6 is chock-full of nifty tips for cutting gobbledygook.)

As a technical writer myself, I never commit one word to my computer until I fill out a Technical Brief such as the one in this chapter and on the Cheat Sheet at the front of this book. Following the Technical Brief document you find a detailed explanation of each item. Feel free to use the Technical Brief as is or amend it to suit your project: Try it! After the first time, it'll take you a short time and save you hours.

Technical Brief

About the Document

1. Type of document
2. Presentation context
3. Target date for completion

Reader Profile

4. Who are the readers?
   - A. Are they technical, nontechnical, or a combination?
   - B. Are they internal (to your company), external, or both?
   - C. Do you have multi-level readers?
   - D. If so, what percentage are there of each?
   - E. What's their level of computer knowledge, if any?
5. What do the readers *need* to know about the topic?
   - A. What's their level of subject knowledge, if any?
   - B. How do they process information?
   - C. What jobs do they perform?
6. What attitude do the readers have toward the subject?

Key Issues

7. What are the key issues to convey? (**A** is the most important.)
   - A. ______
   - B. ______
   - C. ______
   - D. ______
   - E. ______

Project Team

8. Who's who on the project team?

   ______

   ______

   ______

   ______

9. What's the approval cycle? (**A** is the final reviewer.)
   - A. ______
   - B. ______
   - C. ______
   - D. ______
   - E. ______

Executional Considerations

10. Will this document be paper, electronic, or a combination?

    If electronic:
    - A. Modem speeds ______
    - B. Graphic cards ______
    - C. Monitors: Resolution; color or black and white ______
    - D. Offbeat browser or old version of popular browser ______
    - E. The readability on a PC or Mac ______
    - F. Budget constraints ______

    If paper:
    - A. Quantity ______
    - B. Professional printer or computer-generated ______
    - C. Binding ______
    - D. Finished page size ______
    - E. Budget constraints ______

Slicing and Dicing the Technical Brief

Following are the questions on the Technical Brief broken out in detail. After you fill it out once, it's a piece of cake from then on. Your purpose now is to plan your document, not to begin writing it. You see the Technical Brief icon as you go through other chapters. It refers you to specific items on the Technical Brief that help you along your technical writing journey.

About the Document

1. **Type of document:** What type of technical document is this? User manual, product description, reference manual, white paper, abstract, spec sheet, presentation, article, report, computer-based training, Web-based training, or something else?

2. **Presentation context:** If this is part of a project that has other components (such as training, supporting literature, spec sheet, or others), please indicate what they are.

   - If this writing project has no other components, forget No. 2 and go to No. 3. If it does, answer these questions.
   - If this is part of a larger project, can you coordinate efforts with the other team?
   - Are there aspects of either project that can be done in tandem?
   - Can you share any information?
   - Have design or style issues been identified?

3. **Target date for completion:** What is the real drop-dead date? (Too often people pick an arbitrary date, then everyone gives up eating and sleeping to meet it. They later find out the deadline wasn't a real one.) Once you determine the date the document is due, work backwards to fill out the production schedule found in Chapter 3.

Reader Profile

4. **Who are the readers?** Identify your relationship with your readers. Do you share similar experiences and educational backgrounds? Are they familiar with your product or industry?

   For example, a small detail such as knowing their approximate ages may be important in deciding whether to use printed or electronic documents. Surveys show that younger people are more apt to use online documentation than older folks. (Some of you may not be too delighted to hear that *older* means over 40.) Part of this may be attributed to vision problems in the older set. Another may be that younger people grew up in the computer age and computers don't intimidate them.

A. **Are they technical, nontechnical, or a combination?** This will help to determine the backgrounds of your readers so that you can use appropriate language and references. For example, do they share a common background with you or with each other?

B. **Are they internal (to your company), external, or both?** For internal documents, identify your readers by name and job function. For external documents, identify categories of readers (managers, engineers, and others).

C. **Do you have multi-level readers?** If you're writing to multiple-level readers, rank them in order of seniority. For example, if your readers are a mixed audience of managers, technoids, and salespeople, consider dealing with each group separately in clearly identified sections of your document. (Also, determine any language problems.) Here's how you may want to structure the elements in a report for multiple readers:

   - **Table of contents:** Creates a pathway for everyone.
   - **Executive summary (or abstract):** Designed for the managerial level — those who want the big picture only. Find out more about abstracts in Chapter 9 and executive summaries in Chapter 13.
   - **Body:** The main text of the document is for the salespeople, who may represent the majority of readers. The body goes into more detail than the executive summary, but not into as much detail as the appendix.
   - **Appendix:** Appeals to the technoids — those who want all the nitty-gritty details, including data tables.

D. **If so, what percentage are there of each?** This information is critical to help you structure the document. Here's an example:

A number of years ago, I wrote a user manual for a very complex software program. While filling out the Technical Brief, I realized that 20 percent of the readers were hard-core engineers and 80 percent were data entry folks. The challenge was to accommodate such a diverse group of readers. Check out Chapter 8 to see how masterfully I handled that — if I may say so myself.

E. **What's their level of computer knowledge, if any?** For example, if you're writing documentation for a software application, you must know the readers' level of computer knowledge. Following are a few examples:

Greenhorns have little or no computer knowledge. They're prime candidates for the heft and feel of the printed page, not the electronic page.

Sporadic users may have used the system, but they don't use it often enough to remember the commands and other good stuff. They may be amenable to online documentation, if it's easy to use. Or a combination of paper-based and electronic documents may be appropriate.

Aces are the true power users. They understand the ins and outs of the product but may have occasional questions. Their manuals may prop up their screens, and they're prime candidates for online documentation.

5. **What do the readers *need* to know about the topic?**

   A. **What's their level of the subject knowledge, if any?** Think of what your readers *need to know* — not what they already know. You don't want to give too much or too little information.

      - What's their level of knowledge about the subject?
      - What acronyms, initials, or abbreviations will you need to explain?
      - Do they have any preconceived ideas?
      - What are the barriers to their understanding?
      - Is there anything about their style of dealing with situations that should drive your tone or content? Chapter 6 has a full discussion about using the proper tone.

   B. **How do they process information?** During my years of experience in the field, I discovered that people with academic, scientific, or technical backgrounds tend to be process oriented. They benefit from step-by-step explanations. Those with backgrounds in business or law are answer oriented. They respond to quick answers. Creative types are usually visually oriented and benefit from charts, tables, and any visual representation. Discover more about preparing visuals in Chapter 5.

   C. **What jobs do they perform?** Are your readers CEOs, managers, engineers, administrative assistants, data entry specialists, or shop-floor personnel? For example, someone on a shop floor needs hard-copy instructions because he wouldn't necessarily have ready access to a computer.

      You determine the point of view by understanding the needs of your audience. Managers, for example, are big-picture people. They want to know the key issues. Technical folks want the details.

6. **What attitude do the readers have toward the subject?** You may not always tell your readers what they want to hear, but you must always tell them what they need to know. Your readers' attitudes fit into one of these three categories. (You also see an example of what may invoke the reaction.)

    - **Favorable:** You anticipate that the project will be completed one month early. (Who wouldn't be happy to hear that?)
    - **Neutral:** You suggest that everyone stay the course.
    - **Opposing:** You recommend to the CFO that he needs to hire ten more engineers.

Key Issues

7. **What are the key issues to convey?** Every document has a purpose and key issues to convey. If your readers forget everything else, what's the one key issue you want them to remember? (This is akin to putting on an advertising cap to prepare a 10-second spot commercial.) Then proceed to the second most important and so on. Following are examples of points conveyed in order of priority, with **A** being the most important issue.

    A. The medical world's largest, most comprehensive electronic information source

    B. All major medical information in one place

    C. Lowest installation and maintenance fees of all similar services

Turn your key issues into compelling headlines. Chapter 5 explains how.

Project Team

8. **Who's who on the project team?** List all the people involved in the project: subject matter experts (SMEs), writers, editors, graphics folks, and any other "worker bees."

9. **What's the approval cycle?** Consider everyone who'll pick up a pencil and mark up your pearly words of wisdom. Don't be discouraged when this happens. If people don't put marks on the page, they don't feel that they did their jobs. It's nothing personal.

    The final reviewer will probably have minor changes because the document has been through many review cycles. If the changes are major, you may want to start checking the help wanted ads.

Executional Considerations

10. Will this document be paper, electronic, or a combination?

If electronic: Are there technology restraints that affect the design? You must consider that people have a variety of technologies — some state of the art, some state of the art 100 years ago. It's wise to test your electronic document on a variety of technologies (within reason).

A. Modem speeds: Modems don't transmit data as fast as DSL, cable, or ISDN connections. People get frustrated with long downloads on slow modems.

B. Graphic cards: Will the user need a special graphic card?

C. Monitors: Not everyone has modern equipment. Believe it or not, some people still have a black-and-white monitor.

D. Browsers: An offbeat browser or an old version of a popular browser may not display text and graphics properly or at all.

E. Readability: This can be different on PCs and Macs.

F. Budget constraints: Designing for this technology doesn't come cheap, especially when you outsource.

If paper:

A. Quantity: How many copies are needed? The shelf life of the document may determine this number.

B. Printing: Will this be done by a professional printer or generated on a computer?

C. Binding: How will the document be bound? For example, if the document will be inserted into a three-ring binder, you must have the holes punched in the pages.

D. Finished page size: A standard sheet of paper is 8½ x 11 inches. However, finished page sizes vary, so don't take this standard size for granted.

E. Budget constraints: These may determine many of the items just mentioned.

Getting to know your reader

Gathering information about your readers doesn't have to be a daunting task. More and more industries collect speculative information about readers when they plan a product. If you don't have that information, following are ways to gather it:

Before you write

Try to get as much information as you can to reflect the needs of your intended users and the problems they may have. Following are some tips:

- Get some test subjects (guinea pigs) and give them a real-life assignment. Don't give the testers any documentation. Instead, provide a subject matter expert (SME). Have the SME make notes of the questions the testers ask and what they can figure out on their own.
- Do a needs analysis by sending a survey to a broad sampling of intended users.
- Hold training classes before an alpha test. Listen to the questions the students ask and how frequently they're asked.
- Interview intended users. For example, if you're documenting software for the classroom, talk to teachers or observe them in the classroom.
- Look at what the competition is doing.

After you write

You also need to collect some information after your project is completed. Doing so can help you with rewrites of your document. Follow these tips:

- Get feedback from alpha and beta sites.
- Send an evaluation form to get the scoop on what users really think. Check out Chapter 11 for tips on writing questionnaires that get results.
- Get input from the folks on the help lines. Compile a list of the 25 most frequently asked questions.

Part II
The Write Stuff

"The engineers lived on Jolt and cheese sticks putting this product together, but if you wanted to just use 'cola and cheese sticks' in the Users Documentation, that's okay too. We're pretty loose around here."

In this part . . .

Whether you write about an electronic tennis racket that helps you win championship matches, a turbo bra that features a global positioning system, or a corkscrew that tells you when it's time to chill your wine, you must plan the document, write with clarity, provide visual impact, use the right tone, and be accurate.

This part addresses all these issues (not the racket, bra, and corkscrew, but planning, clarity, visuals, tone, and accuracy).

Chapter 3

Creating a Team and a Plan

In This Chapter

- Working collaboratively
- Choosing the right medium
- Creating a production schedule
- Brainstorming and outlining
- Conducting research

There is no reason anyone would want a computer in their home.

—Ken Olsen, president and founder of Digital Equipment Corp. (1977)

In this fast-changing, market-driven, high-pressure world, roles and responsibilities constantly change. You may work with a team on one project, and another team on the next project. The one certainty, however, is that you won't write technical documents in a vacuum; you'll be part of a collaborative effort. A collaborative effort doesn't necessarily mean a large team of people. Your team may be small and consist of a technical writer and a subject matter expert (SME) or a technical writer and a reviewer.

Large teams, however, may include a writer with a team of scientists, engineers, software developers, systems analysts, marketing and sales managers, financial wizards, and the like. Or your team may include other writers, SMEs, editors, publication/electronic specialists, design specialists, and end users. Any combination of these people (or others) creates a collaborative team. When possible, the entire team should be part of the planning process.

Benefiting from the Team Experience

You often find yourself teamed up with diverse groups of people. Be prepared to incorporate their opinions, skills, and styles. Some people enjoy being part of a team; others don't. If you're on the don't side, try to view the experience as an opportunity to benefit from the wisdom and talents of others. You can't control the marketplace, you can't control the weather, and you can't control those higher up the food chain. But you can control the way you relate to team members.

Kicking Off with the Technical Brief

Fill out the Technical Brief (see Chapter 2) when the entire team is assembled so you have everyone's buy-in. Too often, the people involved in the project (even those responsible for signing off) don't get involved at the beginning. If everyone isn't on the same page from the start (pardon the pun), the results can be time delays and missed deadlines.

If you can't assemble the entire team, fill out the Technical Brief with as many people as you can round up. Before going further, get approval from those who aren't present.

Who's on first?

All team members need a clear understanding of their roles and the pecking order. Although this sounds basic, it's often the area where people trip over each other.

To simplify the process, prepare a "Who's Doing What" list to identify the tasks, leader(s), responsibilities, and procedures. Feel free to pick and choose from the checklist shown in Example 3-1. Once the team agrees to each one's responsibility, each member should sign the agreement and receive a copy.

WHO'S DOING WHAT CHECKLIST

Team Responsibilities

- What specific tasks must be completed to finish the project?
- Who will be responsible for each task?

Working Procedures

- When will the team meet?
- Where will the team meet?
- What procedures will be followed in the meetings?
- How will decisions be made (majority or consensus?)
- How will team members communicate? (e-mail? regular mail? one-on-one?)

Example 3-1: Team checklist of who's doing what.

Put your ego aside

Two heads may be better than one, but two egos are worse. Typically, everyone who reviews a document feels compelled to comment. Doing so is just human nature. These comments aren't necessarily a reflection on your writing or your style; they're just part of the process. Never take others' feedback personally.

Turning stumbling blocks into stepping stones

Viewpoints will vary, and conflicts will occur. These differences are a natural part of the collaborative process. Most conflicts, such as grammatical points, are minor. Others, such as basic approaches to a project, are more significant. View these issues as stepping stones, not stumbling blocks.

Team members must find ways to work through all conflicts and make compromises. Doing so is all part of professional growth and helps you produce a top-notch document.

When you challenge a member's viewpoint, do it tactfully and offer a valid reason. For example, if you're writing a major report, you may say, "Your point is well taken, but do you think we should consider putting the key issue first?"

Choosing the Medium That's Right for Your Readers

Once you know the needs of your readers by filling in the Technical Brief in Chapter 2, you can better understand the media they need. To make the decision whether to prepare paper documents, electronic documents, or a combination, ask yourself the following questions:

- **Are the readers computer savvy?** If not, paper is the only way to go.
- **Do the readers have access to computers/or and the Web?** For example, if you're writing a user manual for people who work on a shop floor, you need to know whether they have access to computers.
- **Will you need to update the information regularly?** For example, if you need to let your sales force know of on-the-spot product updates or price changes, you can do that electronically.

Table 3-1 shows the advantages and disadvantages of print documents. Table 3-2 shows the advantages and disadvantages of Web documents.

Table 3-1 Traditional Print Documents

Advantages	*Disadvantages*
Everyone can read paper documents, even people who aren't Web savvy.	Make just one typographical error, and it's there for all the world to see until the next printing.
People are familiar with paper and like its heft and feel.	It's expensive and time consuming to update documents. Commercial printing is labor intensive and costly.
You can pass documents around and share them with colleagues.	You need to provide physical storage space.

Table 3-2 Web Documents

Advantages	*Disadvantages*
Web documents are timely and interactive.	You lock out a percentage of readers because not everyone is Web literate or has access to a computer.
You can provide links to other sites of interest to your readers.	Monitor size impacts your readers' abilities to view your document.
You can update documents on an as-needed basis.	When the computers are down, documents aren't available.
You can provide as many pages as you need because you're not limited by space.	Navigation may be a problem if the site isn't designed properly.

Letting the Production Process Begin

If the team is to meet its deadlines, everyone must have a copy of the production schedule as early as possible. Team members don't always realize the impact of missing a target date, so you may have to stress that point constantly. Although building in additional time is a good idea, it's helpful to include a column for both the target date and actual date. Then you can keep track of a slip in the schedule that endangers the delivery date.

The production schedule is a "must"

Example 3-2 shows a production schedule (also known as a milestone chart) that you may want to use as is or amend for the needs of your project. I always find it helpful to plan the project backwards. Even though the production schedule is listed in chronological order, once you identify the delivery date, you'll know how much time you have for each milestone.

PRODUCTION SCHEDULE
[Name of Project]

Milestone	Target Date	Actual Date	Person Responsible
Fill Out Technical Brief			
Brainstorming Session			
First Draft Delivered			
Prepare Visuals			
Comments Due			
Final Draft Submitted			
Final Visuals Submitted			
Final Draft Approval			
Final Visuals Approved			
Mechanicals (if reqd.)			
Proofs			
Print			
Deliverable/Available	9/16		

Example 3-2: Production schedule listing each milestone.

Check out Chapter 5 to find out how to prepare a production schedule on a *Gantt chart* — a bar graph used to display the time required for activities in a project.

Timing is everything

The amount of time it takes to create a document varies among organizations and the people involved. For example, if key people aren't involved from the get-go, you may have to go back to the drawing board at a later stage in the process to incorporate their new viewpoints. Starting all over chews up a lot of time. To understand the process, analyze historical information on projects the organization has done in the past, and let past experiences be a measuring stick. Table 3-3 gives you some guidelines for timing issues.

Table 3-3 Estimating Your Writing Time

Length of Document	*Original Writing*	*Rewriting*
Less than 100 pages	50 to 100 hours	25 to 60 hours
More than 100 pages	100 to 200 hours	60 to 100 hours

You and others can work on certain phases of the project simultaneously to save time. For example, you can write the text while the graphics people prepare the artwork.

Amassing the Brain Power

Brainstorming, as you see in Example 3-3, is the process of moving ideas from your heads to paper. When the entire team holds a brainstorming session, it provides an opportunity for everyone to give and get input. Someone needs to facilitate, and often the technical writer takes the lead. Following are a few brainstorming tips:

- **Start by drawing a circle.** In the center of the circle, write the purpose of your project.
- **Draw branches and twigs extending from the circle.** Write your main ideas on the branches and your sub-ideas on the twigs.
- **Welcome input from everyone.** Don't pass judgment. This is only a brainstorming session, and any idea is worth capturing.
- **Don't dwell on any one idea.** The purpose of brainstorming is to get as much information as possible.

After you brainstorm, use the twigs and branches to create your outline.

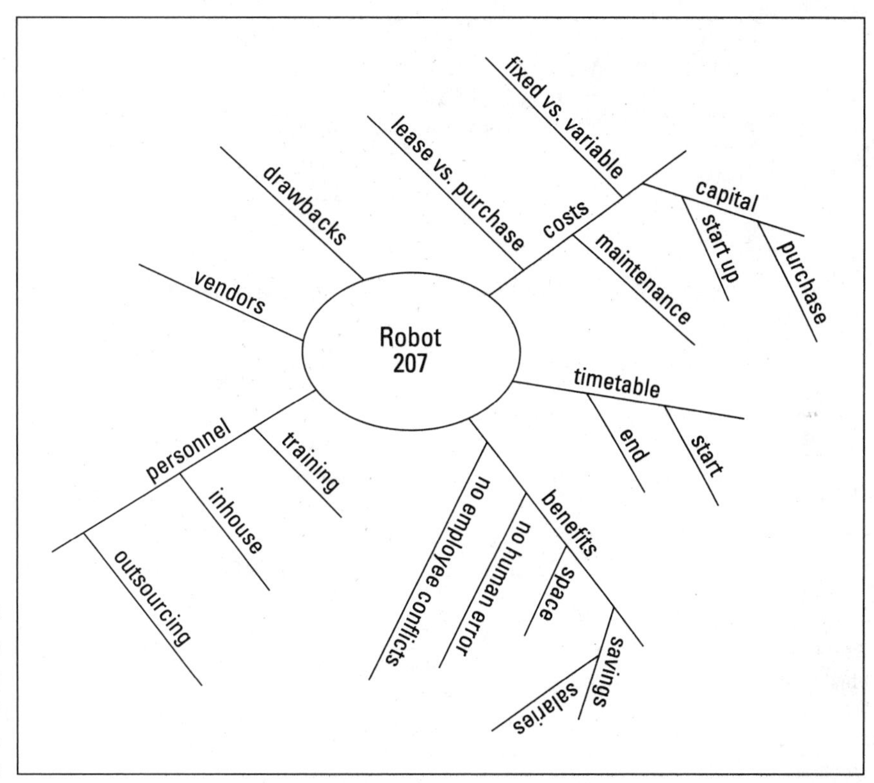

Example 3-3: Brainstorming on how to build a robot.

Generating an Outline

If time allows, generate the outline as a group activity. If not, this activity may fall on one person or a small team. When you complete the outline, get written approval from the key people listed on your Technical Brief under project team.

Writing a traditional outline

Outlining is a tried and tested method from high school. Some people love it; others hate it. Your computer software's outlining feature makes it easy for you to experiment with the organization and scope of information. Example 3-4 is a standard outline.

I. Create Message

A. Preparing a Message

1. Accessing the Create Message Menu
   a. Introduction
   b. Prerequisite
   c. Procedure
   d. Setting the Delivery Default
2. Functions of the Create Message Screen
   a. Forwarding a Message
   b. Changing the Message Type
   c. Verifying the Address

B. Address Defaults

1. Accessing the Address Defaults Screen
   a. Introduction
   b. Prerequisite
   c. Procedure
2. Functions of the Address Defaults Screen
   a. Forwarding a Message
   b. Changing the Message Type
   c. Verifying the Address

II. Online Communication

A. Communicating with IMS

1. (and so forth)

Example 3-4: Traditional outline for a user manual.

Using a decimal numbering system

The decimal numbering system outline shown in Example 3-5 is helpful for documents in which you may need to reference certain sections or paragraphs. If you use this system, you can say to a group, "Look at No. 2.1.2." They'll know exactly where to find the information.

Creating an annotated table of contents

Another outlining option is to create an annotated table of contents, such as the one you see in the front of this book that breaks out the sections in the chapter.

1.0 **BACKGROUND**
  1.1 PURPOSE
  1.2 APPROACH
    1.2.1 Overview
    1.2.2 Tasks
    1.2.3 Critical Concerns
2.0 **TECHNICAL CONCERNS**
  2.1 SUCCESS FACTORS
    2.1.1 Locating Expertise
    2.1.2 Developing Training Procedures
  2.2
  (AND SO FORTH)

Example 3-5: Using the decimal numbering system outline.

Getting Your Arms around the Document

The purpose of writing a technical document is to explain or report on a technical or complex subject. Therefore, unless you're the technical guru writing about something you know intimately, you must research the subject. Gathering data — learning all you can about the product or service — is the lifeblood of the project.

Internal research

If your research is internal, gather your information by interviewing SMEs and others involved in the project, observing people in and around the workplace, reading spec sheets and company literature, and just snooping around. Here are some tips for keeping your nose to the grindstone without getting it cut off:

- **Get to know the product or service intimately.**
  - Test it. Use it. Hug it. Learn all you can.
  - Study system specs and engineering drawings.
  - Interview the SME. Bring a list of questions you prepared.
- **Absorb the big picture.**
  - Find out about the purpose of the produce or service, who'll use it, why, and when.
  - Learn the acronyms.

   - Become familiar with the good and bad features of the product or service.
   - Learn about any possible flaws with the product, such as what features may break, malfunction, or cause problems.
- **Cozy up to the marketing and sales departments.**
   - Learn the advantages and disadvantages of the product or service.
   - Understand the features and benefits you need to highlight.
   - Know how the manual will be distributed and updated.
   - Find out whether there are related publications or manuals.
- **Find out whether you need to prepare unpacking instructions or a parts lists.**
   - Ask what's in the shipping carton.
   - Ask whether the user needs instructions to install or assemble the item.
- **Gather as much written or graphical data as you can. You can always discard what you don't need.**
   - Decide what photos or graphics you need.
   - Get copies of a style manual or standard text that similar projects used.

It's important that you don't merely trust your memory. Take copious notes while doing your research.

External research

You do external research by interviewing people outside your organization, visiting the library, or surfing the Web. Here's some advice to make your job easier:

- **Interview people outside your company.** You can either visit people personally or send a questionnaire. Which to do, of course, depends on the nature of your research. If you need to gather information from a lot of people, you may find a questionnaire is appropriate. Check out Chapter 11 for tips on preparing a questionnaire that evokes responses. If you need to meet face to face with one or more people, follow these guidelines:
   - Prepare a list of objectives and questions beforehand.
   - Let the interviewee do most of the talking. After all, you're there to learn.

  - Grab all the literature the interviewee is willing to share.
  - Take copious notes or use a tape recorder, if appropriate.

- **Do research at the library.** There are a variety of ways to use the library for research. If you need help using any of the following library services, ask the librarian:
  - **Card catalog:** Card catalogs are alphabetically arranged by author, title, and subject to reflect all the books and resources the library owns.
  - **Bibliographies and periodical indexes:** Bibliographies list periodicals, books, and other published research in a wide variety of subject areas. Periodical indexes list magazines, journals, and newspaper articles. Libraries sometimes house bibliographies and periodicals in a separate room. Newer indexes may be on the shelf, while older indexes may be in another part of the library or in microform.
  - **Reference works:** These include encyclopedias, dictionaries, handbooks, manuals, statistical sources, atlases, and more.
- **Surf the Web.** If you don't know the URL of the site you want, access one of the popular search engines. Check out Chapter 14 for in-depth tips on using the Web for research. Following are a few highlights:
  - **Type in a keyword or phrase.** The search engine looks for documents that have your keyword(s). It then ranks the sites based on how many times your keyword appears in the document, whether the keyword appears in the title, how early the keyword appears in the text, and so on.
  - **Try to make your search as narrow as possible.** Otherwise, you may get thousands of hits. For example, if you search for *dog,* you get Web sites for more dogs than you want barking at you. If you search for *poodle,* you get all types of poodles. If you search for *toy poodle,* you narrow the search to the toy variety only.

Chapter 4

Don't Be a Draft Dodger

In This Chapter

- Putting meat on the bones
- Dealing with rounds of edits

I have traveled the length and breadth of this country, and have talked with the best people in business administration. I can assure you on the highest authority that data processing is a fad that won't last out the year.

—Business editor at Prentice-Hall, 1957

You may be wondering why this chapter is so short. When you use the Technical Brief (discussed in Chapter 2) and you plan properly (as Chapter 3 explains), you've provided the structure to draft your document with ease. With proper planning, writing the draft is just like adding meat to the bones. If you don't plan properly, you may sit and sit and stare at your computer hoping that pearly words appear on the screen.

Psyching Yourself Up

The most difficult part of writing the draft is getting started. Therefore, don't wait for inspiration; just sit down and do it. Following are some helpful hints to put you in the "write" frame of mind:

- **Get comfy.** Try to get as comfortable as you can because comfort enhances concentration. In today's cubicle settings, creating the environment you want isn't always easy, but you can do a few things for yourself. If you like to air your feet, slip your shoes off. If you like to munch, get a bag of tortilla chips or carrot sticks. If you like music, put on a headset.
- **Gather what you need.** Gather all your notes and reference materials before you start. When you constantly get up to find what you need, you break your train of thought.

- **Set reasonable time frames.** Set a reasonable period of time for each segment of your writing. You may set a goal to write for a half hour or an hour. Continue to write until you meet your goal even if the words aren't flowing. If you're on a hot streak, by all means keep chugging along.

Getting Down to Business

This is where all your planning pays off. You already have the outline (in whatever form you generated it); therefore, writing the draft is akin to filling in the blanks. Here's how to proceed:

- **Write one section at a time.** Start with the one that's the easiest to write. Your readers will never know where you started.
- **Avoid the temptation to go over what you write.** The important thing is to keep moving forward.
- **Don't worry about spelling, grammar, and punctuation.** You worry about those issues later in the process when you proofread.
- **Keep moving forward.** If you can't think of the right word, use another and keep going. This isn't a finished document — it's your first offer.

After you finish each writing session, get some distance. Even if you're on a tight deadline, take a five- or ten-minute break. Put your feet up. Go for a brief walk. Return a telephone call. Get a cup of coffee. Pat yourself on the back. You need to clear your head and refresh your brain.

After you write the draft, incorporate the guidelines in Chapters 5, 6, and 7 before you send your documents out for review.

- **Chapter 5 talks about how to make your documents visually appealing.** Your documents need strong visual impact in order for people to want to read them.
- **Chapter 6 discusses honing the tone — how you "sound" to your readers.** Your words tell your readers a lot about you because they create a window to your mind. You always want to "talk" to your readers in a tone that's appropriate for them.
- **Chapter 7 explains ways to proofread effectively.** It also has a great checklist to use for all your technical documents.

Integrating the Editing Process

Rarely will you write a document without getting input from others — before, during, and after the writing process. In a long technical document, you may have several rounds of revisions. Before you circulate each revision, check out Chapter 7 for tips on editing and proofreading. You don't want your draft to make its rounds with errors.

Brown-paper editing

Here's a nifty process that has less to do with the color of the paper than with the process. I worked with a consulting company where people made edits on a draft of the document taped to a wall lined with brown paper. The writers found a large expanse of wall space in an unused conference room or hallway and used that space for editing documents. Here's how the process works — and I was amazed at how well it works!

1. **Buy rolls of brown paper and tape.**

   Purchase several rolls of brown paper (such as the kind used to wrap packages for mailing) and a roll of quick-release tape that doesn't damage your walls when you yank it off.

2. **Tape the brown paper horizontally across the walls.**

   Be sure to provide enough wall space to tape your document. You can do this in several layers so the paper isn't too high or too low for people to read and write on.

3. **Print your document, double-spaced, and tack each page to the brown paper in sequential order.**

   You may do this one chapter at a time or whatever works for the length of your document. You may also do this for each round of revisions.

4. **Schedule a date and time for each person or group of people to edit the document.**

   Provide a different color marker for each person or group so you know who wrote what on the draft of the document.

This process works well because it makes people commit to a time for this task or they lose their right to comment. This process also prevents the all-too-common situation in which a document sits in someone's inbox until the pages turn yellow.

All you have to do as the writer is post the document on a wall, generate a schedule for people to edit, and make sure they know the time and place. Build this process into the production schedule and people know you mean business. Check out Chapter 3 for more information about creating a production schedule.

Editing electronically

When several writers work on a draft, they should learn how to annotate text without destroying the original. Following are several techniques you may try:

- Consider using the strikethrough feature or a different typeface or color.
- Use proofreading marks if they're available in your software.
- If you use an updated version of Microsoft Word, leave the tracking marks by accessing Tools➪Track Changes➪Highlight Changes and checking the Track Changes While Editing box.

Holding on to your ego

Everyone who reviews the document will undoubtedly make at least one change. When people pick up a pen or edit on the computer, they feel compelled to change something. *Don't take the changes personally;* they're the nature of the beast. Ask the members of the team to review and edit the draft carefully and critically. Remind them to check for large and small issues — from the organization of each segment to the technical accuracy. If the edits are extensive or differ dramatically from what the team planned, give team members the option of accepting or rejecting the changes.

Major changes may lead to another round of planning. That's not a negative result. It's just another opportunity to evaluate the document and make it better. After all, until the time a document goes out the door, it's a work in progress.

When I review a document, I never use a red pen — it's too much like the dreaded marks your teachers made on your school papers. I generally use a green pen. Green is an organic color that's less bloodlike.

Revise and Consent

Once you have all the edits, decide which changes to incorporate into the document. (Reviewers generally get a second pass.) This time the edits should be minor and easy to include.

Before your document is ready to be printed or go online, check out Chapter 7 and proofread, proofread, proofread. However, you must know when to stop. Even the most wonderful writers can always make something better. Stop when you notice any of the following:

- **You start nitpicking about insignificant words.** For example, you get wrapped around the axle trying to decide whether you should use *easy* or *simple?* Pick one and get on with your life.
- **You revise the revisions of the revisions of the revisions.** At some point, you start making changes for the sake of making changes. If you're not adding value, give it up.
- **You can't stand to look at the document one more time.** That's the telltale sign to pack it in.

Chapter 5

Visualize This!

In This Chapter

- Using white space
- Optimizing sentence and paragraph length
- Pumping up headlines
- Preparing bulleted and numbered lists
- Sequencing for the reaction of your reader
- Generating charts, tables, and figures
- Using color effectively

For G-d's sake go down to reception and get rid of a lunatic who's down there. He [Logie Baird, inventor of the television] says he's got a machine for seeing by wireless! Watch him — he may have a razor on him.

— Editor of the *Daily Express* (London), 1925

You may have spent days, weeks, or months gathering information and writing a great technical document. If the document doesn't have visual appeal, however, nobody will read it or understand it. That's akin to spending months lining up an interview for your dream job. If you walk into the interviewer's office with a ketchup stain on your lapel, nobody will hire you because you aren't visually appealing. To ensure that your document gets the attention it deserves, it too must be visually appealing.

You don't need to be a graphic artist with fancy software to create a pleasing visual design that has impact on your reader. This chapter walks you through simple ways to turn *ho-hum* documents into *smashing* documents.

This chapter covers the importance of visual impact in paper-based and electronic documents. However, I recommend that you check out Chapter 15 for visuals that are specific to the Web, such as animations, frames, marquees, and more.

May I Have Your Attention, Please?

Visuals serve as attention-getters to communicate information at a glance. They provide a subtle, unconscious signal that the document is worth readers' attention. When a document has visual impact, it attracts attention, invites readership, and establishes the credibility of your message even before you state your case. Here's why:

- **Visual impact organizes information.** A good visual design breaks the document into manageable, bite-sized chunks, making it easy for readers to find the key pieces of information. The readers can concentrate on one idea at a time.
- **Visual impact emphasizes what's important.** You can create a hierarchy of information so that your readers can separate major points from supporting ones — much like you see in newspapers. In today's harried world where people are pulling their hair out because of tight schedules, your readers will appreciate a quick read.

Using White Space

White space is a key ingredient in visual design; it includes all areas on the page or where there's neither type nor graphics. (On a computer screen, this area is referred to as *quiet space* or *blank space.*) White space doesn't have to be white. For example, if your paper or screen background is ivory, tan, or whatever, the background color is called white space. Here's what white space does for your document:

- Makes it inviting and approachable
- Provides contrast and a resting place for the readers' eyes
- Creates the impression that the document is easy to read

Enjoy the white open spaces

Following are tips for using white space effectively:

- For paper documents, use 1- to 1½-inch top, bottom, and side margins to create a visual frame around all the text and graphics. For electronic documents, leave a ¼ to ½ inch margin all around.
- Double-space between paragraphs to help the reader see each paragraph as a separate unit.
- Emphasize key pieces of text (words, phrases, or paragraphs) with white space or a different font.

Give Me a Break

It's crucial to break your sentences and paragraphs into manageable, bite-sized chunks of information. Many technical writers use long sentences and dense paragraphs. Doing so makes technical information difficult to digest and causes readers to tune out. When you optimize sentence and paragraph length, you give your documents more visual appeal.

Limit sentences to 25 words

One way to stick to the 25-word limit is to look for compound sentences — you know, those separated by *and, but,* or other conjunctions. Notice how the following lengthy sentence is chopped into three sentences:

- **Lengthy (39 words):** "As you see in Diagram A, variations across the die arise from stencil aperture dimension variations and stencil cleanliness, and smaller variations arise from random defects such as inclusions in the paste and contamination from the wafer or environment."
- **Just right (broken into three sentences):** "As you see in Diagram A, variations across the die arise from stencil aperture dimension variations and stencil cleanliness. Smaller variations arise from random defects. These may include inclusions in the paste and contamination from the wafer or environment."

Up, up, and oy vey!

In its day, the Boeing 747 was considered an engineering phenomenon. It holds up to 490 passengers and is 2,775 inches long (that's longer than two basketball courts). This 710,000-pound marvel can leap tall buildings in a single bound. The 747 took five years to develop, which included hundreds of thousands of labor hours. Imagine what the documentation was like.

The documentation contained 31,000,000 (yes, that's 31 million!) sheets of instructions, plans, specifications, diagrams, parts, and change orders that included buzzwords, symbols, figures, and more. If every Boeing 747 tried to carry that tome on board, I wonder whether the plane could have taken off. Better yet, imagine keeping those tomes on your bookshelf and poring through them. (Fortunately, there is a better way. With Web technology, a lot of this data is stored in cyberspace and accessed as needed.)

Limit paragraphs to 8 lines

Think of your paragraphs as trains of thought. When one train leaves the station, another train arrives that heads in the same general direction. Although there are no hard and fast rules about paragraph length, when you limit each paragraph to eight lines, you have a very readable document.

- **Dense paragraphs:** When you write dense paragraphs, readers find your text intimidating. They fail to see your subdivision of thoughts.
- **Short, choppy paragraphs:** When you present readers with a lot of short, choppy paragraphs, it's difficult for them to see the logical relationship between your ideas and thoughts.

Harness the Visual Power of Headlines

Newspapers and magazines use informative headlines as guideposts for visual impact. The headlines tell a story and direct the readers to what's important. When you write compelling headlines, readers skim the message, and the headlines tell the story. *As a writer,* you tell your reader what's important and direct the flow of information. *As a reader,* you get the gist of the text and find key information quickly.

Notice how the "Informative" headlines that follow give the reader vital information at a glance:

Informative: Introduction: XYZ Machine Holds Great Promise
Noninformative: Introduction

Informative: Quarterly Inspections Cut Accident Rates by 23%
Noninformative: Report of Quarterly Inspections

Informative: Conclusion: We Need to Conduct Further Tests
Noninformative: Conclusion

Informative: Findings: There Is No Critical Difference Between the Control Group and the Experimental Group
Noninformative: Findings

Put It on the List

If you ever believed in Santa Claus, you know all about making lists. When you prepared your Christmas wish list, you wrote the hottest item as No. 1; the second hottest, No. 2; and so on. If you didn't use numbers, you wouldn't have given Santa any visual clue as to what was important to you. He may have just picked a few things you asked for, and then you'd be disappointed on Christmas morning when your shiny red Jaguar wasn't waiting for you.

When you prepare a shopping list, you list each item but don't use numbers. Once you're in the store, you just pick items off the shelf — everything has the same weight (figuratively speaking).

Following is an explanation of when to use a bulleted list or a numbered list:

Using bulleted lists

Use bulleted lists when rank and sequence aren't important. Bullets give everything on the list equal value. Always head the list with a descriptive sentence, as you see in Example 5-1.

Following are the fabrication methods for stencils:

- Laser cutting
- Chemical etching
- Electroforming

Example 5-1: All bullets are created equal.

Using numbered lists

The bullets that follow show you when to number a list. (Notice that I didn't use numbers for this bulleted list because one item doesn't have priority over another.)

- **Show items in order of priority.** Doing so gives the reader a visual clue that the items on the list are in priority order. (See Example 5-2.)
- **Describe steps in a procedure.** When you describe steps in a procedure, start each numbered item with an *action word* — something for the reader to do. (See Example 5-3.)
- **Quantify items.** If you don't number a long list, people count the listed items in their heads to make sure the number of items is correct. When you number the list, you let readers reserve their brain power for more important things. (See Example 5-4.)

Please take care of these issues first thing in the morning. Thanks.

1. Call the ABC Agency to arrange for a consultant for the week of March 15.
2. Ask Jim to prepare his R&D report.
3. Schedule a meeting with everyone involved in the project for the week I return.

Example 5-2: Putting first things first.

Following are the requirements for paste formulated for wafer printing:

1. Use a squeegee action to deliver all the stencil aperture contents to the UBM surface.
2. Remove any remaining solder beads with the automated wiping process.
3. Remove oxides from the solder beads during the reflow process.
4. Remove flux residues after the reflow process with mild chemistries.

Example 5-3: Take it one step at a time.

Following are the ten key people on the team:

1. Jon Allen
2. Samuel Jones
3. Kim Wong
4. Jackson Pollack
5. Jane Robinson
6. Quincy Adams
7. Dwight Alexander
8. Barbara Geller
9. Pat Lewis
10. Morton Karp

Example 5-4: Don't bother counting.

Writing lists in general

Certain guidelines apply to bulleted and numbered lists. Punctuate properly, use parallel structure, and break lists into manageable chunks of information.

Parallel structure

Imagine a gymnast in the final tryouts for the Olympics. She gracefully dances along the parallel bars; her eyes are aglow as she looks and smiles at the audience. All of a sudden — oops! — the bars aren't parallel. One bar veers to the left. The poor gymnast falls to the floor. Now imagine your readers, totally absorbed in your document. All of a sudden — oops! — the sentence isn't parallel. One component veers off. The poor readers' expectations fall.

Whether you use a bulleted or numbered list, create items that are parallel in structure. That means all elements that function alike must be treated alike. For example, in the parallel bulleted list that follows, all the bulleted items are gerunds — they end with *-ing.* In the nonparallel bulleted list, the first two items end with *-ing,* making the last item stick out like a wart at the end of your nose.

Parallel bulleted list:

Effective measures should involve

- Designing and maintaining the facility
- Training the operators and other people in the field
- Specifying security personnel and procedures

Nonparallel bulleted list:

Effective measures should involve

- Designing and maintaining the facility
- Training the operators and other people in the field
- Specifying security personnel and procedures

Punctuating a list

People often get confused as to when to use a colon to introduce a list and when to use a period to end a list. The following demystifies these pesky marks of punctuation. (For more about punctuation, check out Appendix A.)

- **Colon:** Use a colon to introduce a list when the words *the following* or *as follows* are stated or implied. However, don't use a colon after a verb (as you see in the examples in the section "Parallel structure").
    - Please consider the following ideas:
    - Please consider these ideas: (*The following* is implied.)
    - The three factors are (In this case, don't use a colon. Just follow the sentence with the bulleted or numbered list.)
- **Periods:** Use a period after each item in a list only when the items on the list are complete sentences. When the items complete the sentence (such as in the examples in the section "Parallel structure"), put a period after the last item only.

Avoiding laundry lists

When you have too many items on a list, you create a laundry list and readers may just gloss over everything you worked so hard to emphasize. Instead of creating a long list of bulleted or numbered items, break the items into categories. In Example 5-5, you see one long laundry list. In Example 5-6, you see how dividing the list into two logical chunks of information is easier to read and gives more information.

Our global expansion takes us into the following countries:

- Austria
- China
- Hong Kong
- Indonesia
- Malaysia
- Portugal
- Spain
- Sweden
- Thailand

Example 5-5: Laundry list of bulleted information.

Our global expansion takes us into the following countries:

Asia

- China
- Hong Kong
- Indonesia
- Malaysia
- Thailand

Europe

- Austria
- Portugal
- Spain
- Sweden

Example 5-6: Logical chunks of bulleted information.

What's your sign?

If you want to look savvy in print, use the special signs and that come with your software. If you use (c) for copyright or - - for the em dash, your visual effect will appear amateurish. These "prehistoric" signs date back to the dark ages of typewriters that had limited characters. Following are just a few signs and symbols that are popular in the technical world:

Sign or Symbol	Used for	Sign or Symbol	Used for
©	Copyright	Å	Angstrom
®	Registered trademark	λ	Lambda
™	Trademark	μ	Mu
π	pi	√	Square root

The Natural Order of Things

We can thank the academic world for teaching us how to write. However, writing in the academic world didn't teach us how to write in the business world. Academic writing tends to be long and labored; business writing is (or should be) brief and to the point.

Can you recall writing a report for class? You wrote a long introduction, built a very extensive case that went on at great length, and at the very end wrote your conclusion. Well, in the business world "conclusion" isn't synonymous with *end.* It's synonymous with *findings.* When the findings are important, should they be at the end? Perhaps *yes,* and perhaps *no.* Place key elements based on how your readers will react to your message.

You must sequence information to have the maximum impact on your readers. This is where the Technical Brief in Chapter 2 and on the Cheat Sheet in front of this book is essential. Look at No. 6, "What's my readers' attitude toward the subject?" You must know whether your readers will be favorable, neutral, or opposing to your key issue.

Put good news up front

Everyone likes to be the bearer of good news. Therefore, when readers will be favorable or neutral to your message, put the key issue up front. Why hide it? Put the findings, conclusions, or recommendations at the beginning of the document. Why make readers wade through reams of pages before getting to the key issue?

Writing equations

Even mathematical equations have style. When you write an equation as a sentence, you can give it the form of normal text or break it out on a line of its own.

For example, if you want to use an equation within a sentence, you may express it on one line as follows:

The equation you need to know is $z={}^{1}\!/_{x}=y$

Or you may decide to break it out on a line of its own.

The equation you need to know is

$$z={}^{1}\!/_{x}+y$$

Following is the structure for a report with this scenario: You just completed a study to determine roadblocks to getting a piece of equipment completed in time for the rollout. Based on lots of data you accumulated, you recommended that the company hire five more engineers in order for the project to meet its target date. Do you think the engineers who are currently working 60-hour weeks will welcome the news? Of course they will. When you put your recommendation ahead of the analyses and background, as you see in Example 5-7, these happy employees may mention you in their wills.

Favorable or neutral audience
Title Page
Executive Summary
Table of Contents
Purpose
Recommendation
Analysis and Supporting Data
Background
Summary
Appendices

Example 5-7: Being the bearer of good news.

Delicately present bad news

You may not always tell your readers what they want to hear, but you must tell them what they must hear. For example, using the scenario in the earlier section "Put good news up front," do you think the chief financial officer will welcome the news that she must hire five more engineers in order for the project to meet its target date? Of course not. Her bottom line is negatively impacted by these costs.

Therefore, it would be wise to present the background — the problems that precipitated this study. Then present the analysis you gathered to support the background. After you build your case, present the recommendation. In doing that, you've built the reader up to the "grand finale" instead of hitting her between the eyes with news she may not want to hear. Example 5-8 shows how to structure the same report for an opposing reader.

Example 5-8: Being the bearer of bad news.

Opposing audience
Title Page
Executive Summary
Table of Contents
Purpose
Background
Analysis and Supporting Data
Recommendation
Summary
Appendices

When you write to readers that you expect to have different reactions to your document, sequence for the decision maker. One way is to prepare an executive summary, keeping the reaction of the decision maker in mind. (Check out Chapter 13 for details about writing an executive summary.) Remember that decision makers generally don't want all the facts; they focus on the big picture and may read an executive summary rather than the entire document.

With online documents, there's no fixed order. Readers can click on a hyperlink, and you don't know in which order they'll access information.

Use the sequence that works best

As a writer, you must decide which sequence will have the impact you want to have on your readers. Table 5-1 shows a variety of methods for specific information flows.

Table 5-1 Sequencing for Impact

Information flow	*Method*	*Uses*
Cause and effect	Show a plausible relationship between a situation and its causes or effects.	Experiments, accident reports
Chronological	Arrange events in sequential order to stress the relationship of what happened and when. Begin with the first event and continue to the last.	Trip reports, trouble reports, minutes of meetings, work schedules, manufacturing or scientific procedures, test protocols
Comparison	Point out similarities or differences, or advantages and disadvantages. (Tables or graphs are great ways to present these.)	Feasibility studies, research results, trends and forecasts
Decreasing order of importance	Start with the most important point and end with the least important point.	Reports for decision makers who make decisions based on most important point
Division and classification	Divide complex topics into small chunks of information.	Processes, instructions
General to specific	Begin with a general statement and then provide facts to support it.	Reports, memos
Increasing order of importance	Start with the least important point and end with the most important point.	Personnel goals, oral presentations
Sequential	Explain something step by step.	Instructions, user manuals
Spatial	Describe an item according to the grouping of its physical features. This relates to where things are from east to west, north to south, left to right, top to bottom, interior or exterior.	Activity reports, layout of equipment, building sites, research reports
Specific to general	Start with a specific statement and build to a conclusion. A good tool for persuasive writing.	Analogies, work orders, customer service responses, feasibility reports

A Pixel (Picture) Is Worth a Thousand Words

Charts and graphs are super ways to make your point very effectively. You can gather data and prepare a chart to display your findings, identify opportunities as a result of what visually appears, and update the data to show changes or progress. Many software applications are available to help you prepare graphs in a jiffy. Check the Internet or your local computer to find out more about them.

Keep these tips in mind when you prepare charts and graphs:

- **Write a descriptive title.** Place the title above the chart or graph.
- **Use an appropriate scale.** For example, if your financial range is from $100,000 to $200,000, don't show a scale of $100,000 to $500,000.
- **Create a legend if the chart isn't self-explanatory.** Legends explain the symbols that appear in the chart.
- **Keep the design simple.** Eliminate any information your readers don't need to know.
- **Prepare a separate chart or graph for each point.** If you try to squeeze too much information on one graph, you defeat your purpose of making it simple to read.

If a pixel is truly worth a thousand words, you can eliminate the thousand words with a well-done graphic. Make the graphic self-contained, tie it to the text, and place it as close to the text as possible. Clearly label all the parts so the graphic is self-explanatory and sends a clear message.

Pie chart

A pie chart is like a pizza with wedge-shaped sections. You may order a pizza with 50% pepperoni, 25% mushrooms, and 25% olives. Each section represents a percentage of the total pie, which is 100%.

Some people think it's important to begin the most important percentage at the 12 o'clock position and continue clockwise. Others believe that (because people read from left to right) the most important information should be to the left of 12 o'clock and continue counterclockwise. It's your choice. In Example 5-9 you see what a typical pie chart may look like. Example 5-10 shows a pie chart in three dimensions. (Of course, the 3-D pie has more calories.)

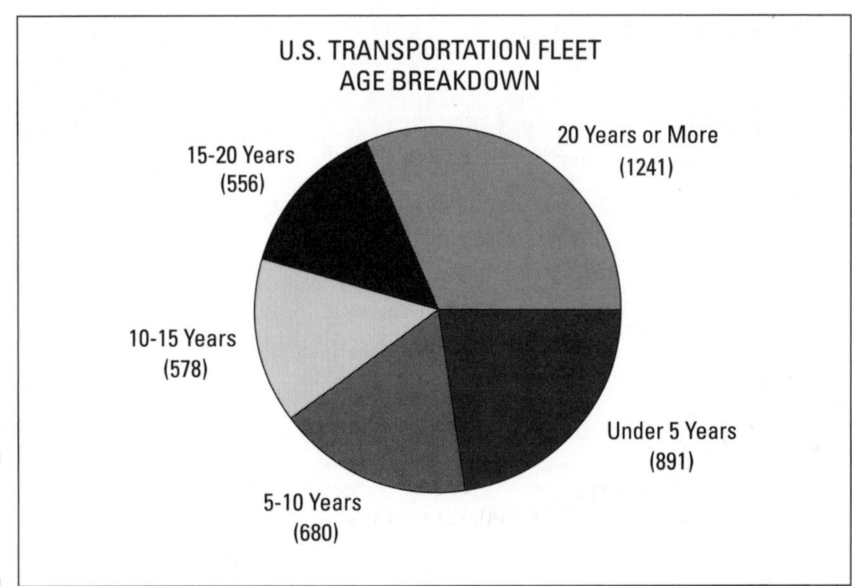

Example 5-9: Here's pie in your eye.

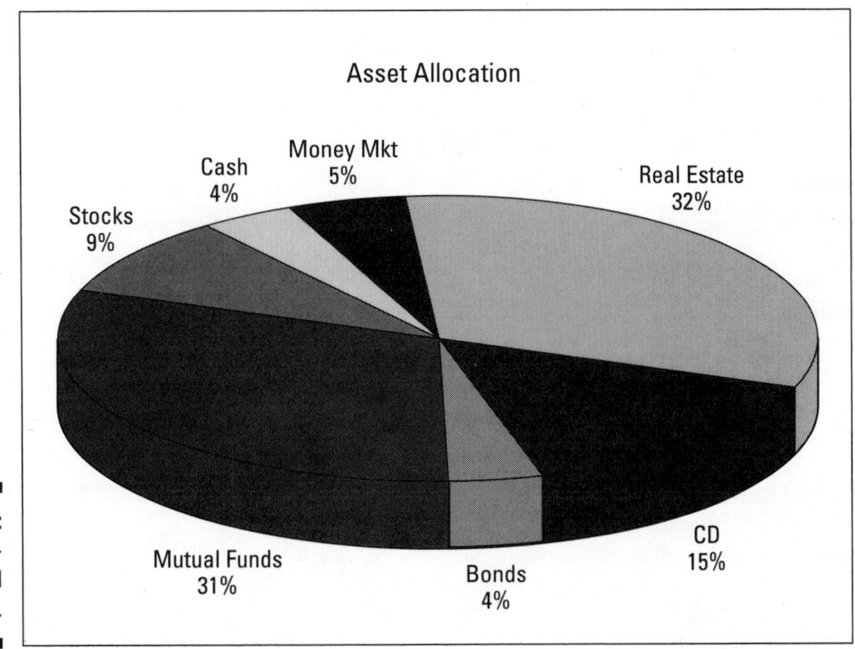

Example 5-10: Three-dimensional pie.

Line chart

A line chart shows trends or the change of one or more variables over time periods, as shown in Example 5-11. Line charts use points plotted in relation to two axes drawn at right angles. Make the axes descriptive and use clear labels.

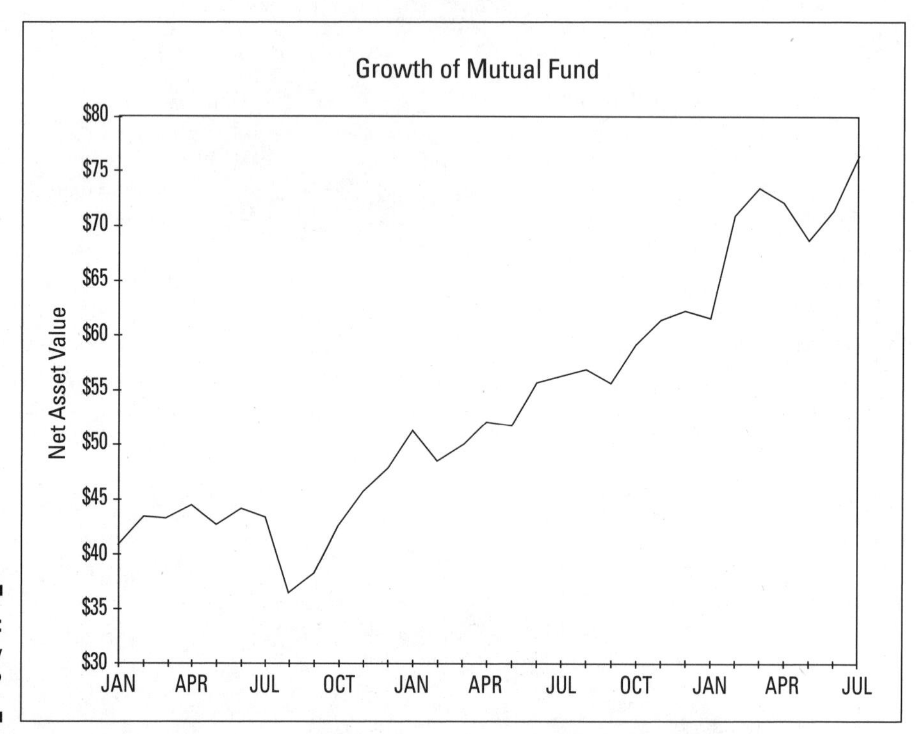

Example 5-11: What's my line?

A slight variation to the line chart is the *run chart* that shows incidents above and below an established data point. This is evident when you compare Example 5-11 with Example 5-12.

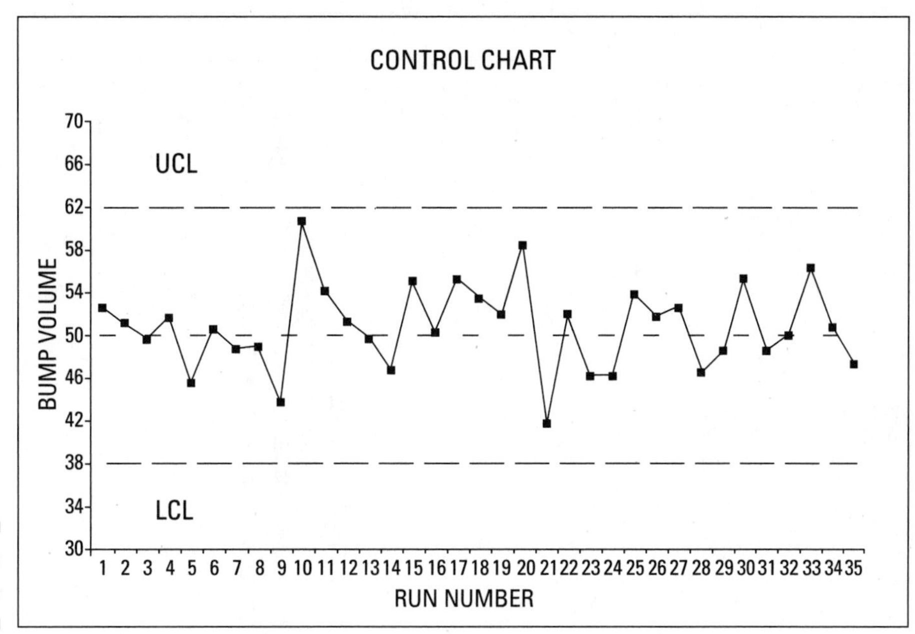

Example 5-12: On the run.

Bar charts

A bar chart (which can be vertical or horizontal bars) shows a comparison between categories, as you see in Example 5-13. Clearly mark the axes. Variations to simple bar charts are histograms, Pareto charts, and Gantt charts, which you use for specific purposes. I describe and give examples of these charts in the following sections:

Histogram

A histogram shows the relative frequency of occurrence, central tendency, and variability of a data set, as you see in Example 5-14.

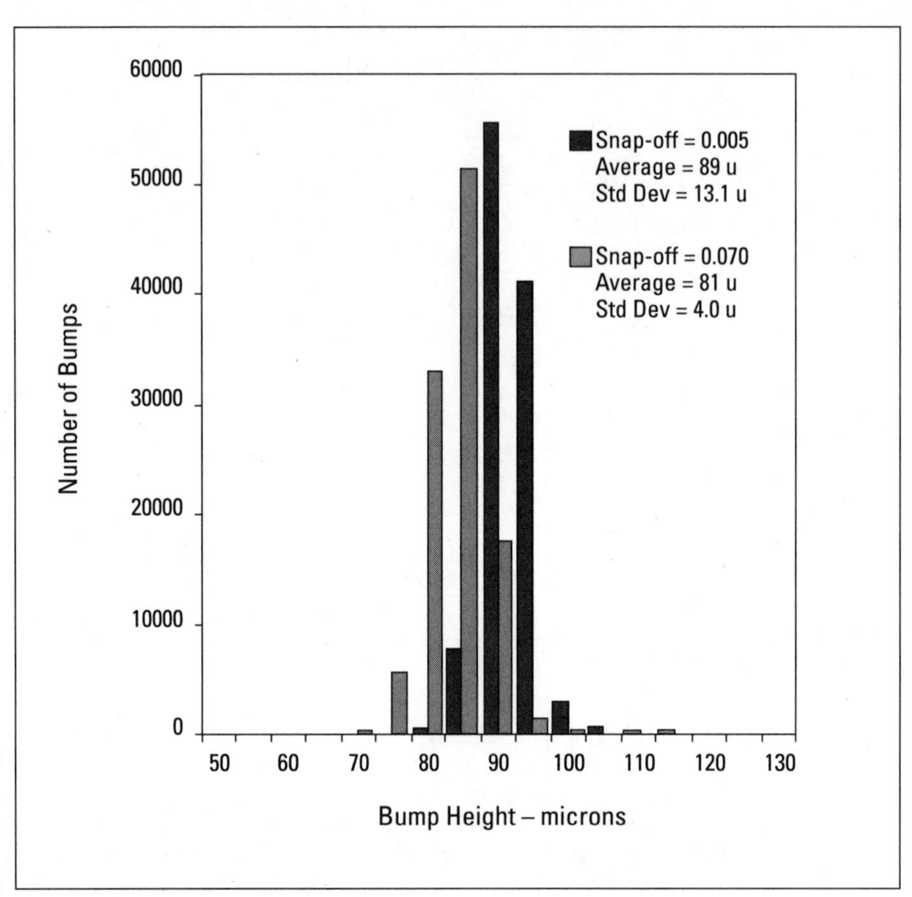

Example 5-13: Step up to the bar.

Pareto chart

A Pareto chart, shown in Example 5-15, separates vital information from the trivial information. It's based on the Pareto Principle, which says that 20 percent of the problems have 80 percent of the impact.

Gantt chart

The Gantt chart, shown in Example 5-16, is a tool used by management to help coordinate resources and activities. It shows timing relationships between the tasks and subtasks of a project.

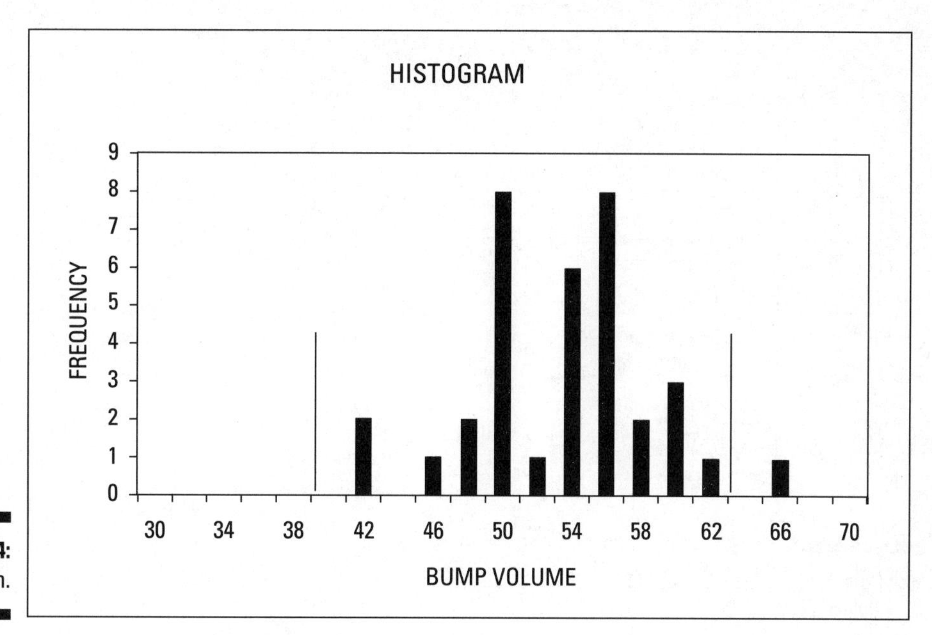

Example 5-14: Histogram.

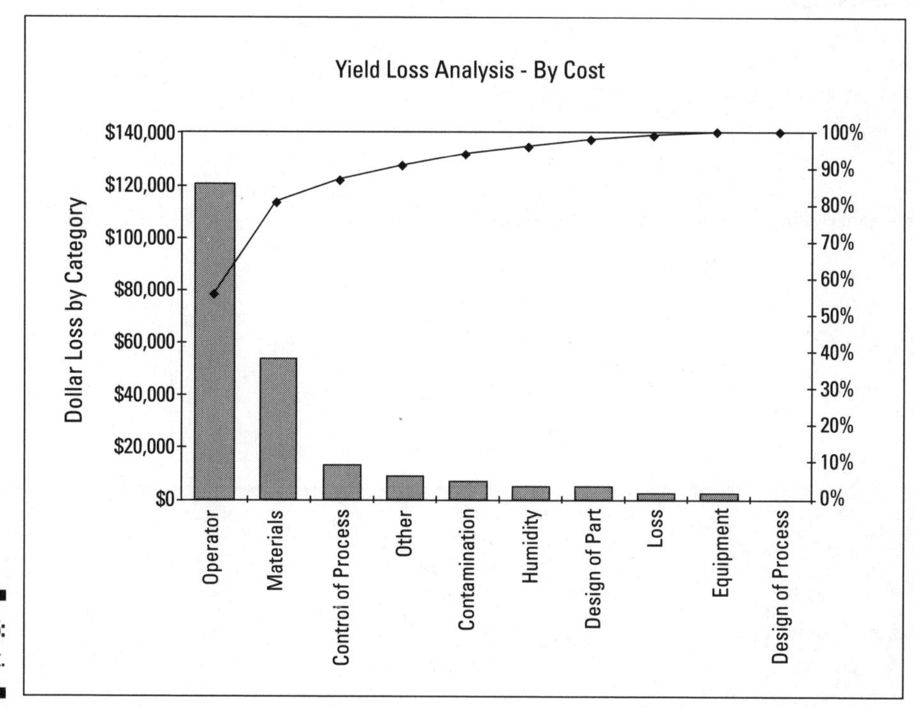

Example 5-15: Pareto chart.

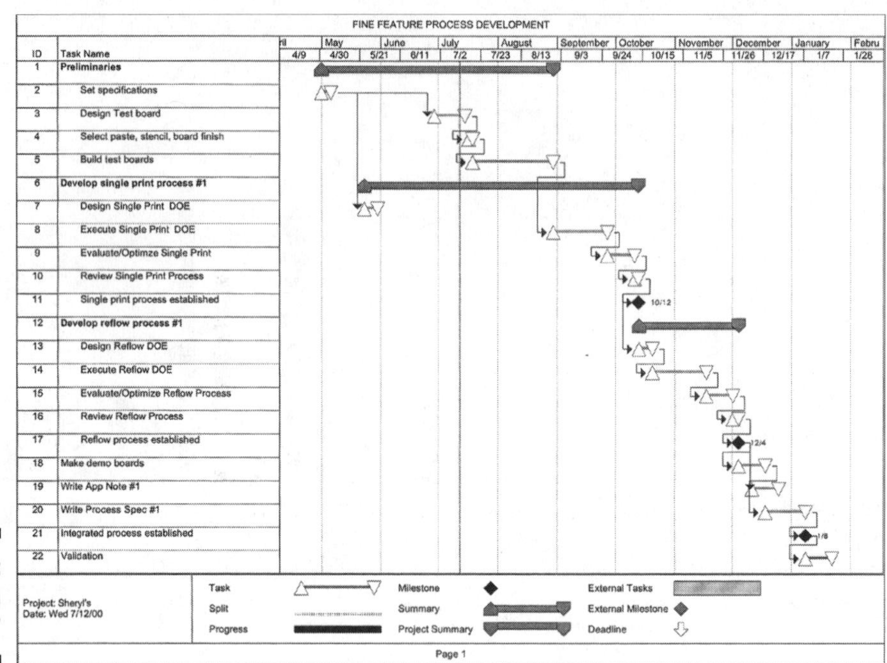

Example 5-16: Who's on first?

Scatter chart

A scatter chart, shown in Example 5-17, displays a relationship between two variables. It may help pinpoint the cause of a problem or show how one variable may relate to another.

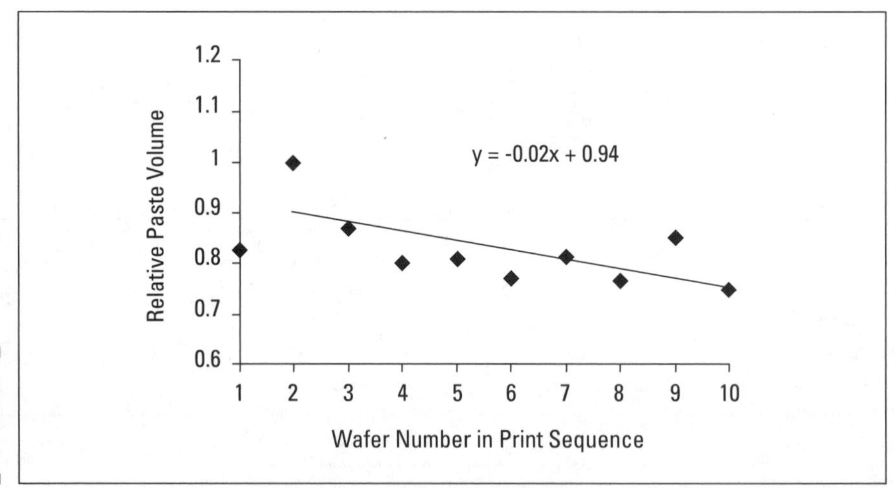

Example 5-17: Scattered around.

Flowchart

Example 5-18 shows symbols used in a flowchart. Example 5-19 displays the major steps in a process using flowchart symbols.

Standard Flowchart Symbols

This symbol...	Represents...
(rounded rectangle)	Start/Stop
(diamond)	Decision Point
(rectangle)	Activity
(document shape)	Document
(circle)	Connector (to another page or part of the diagram)

Example 5-18: Go with the flow.

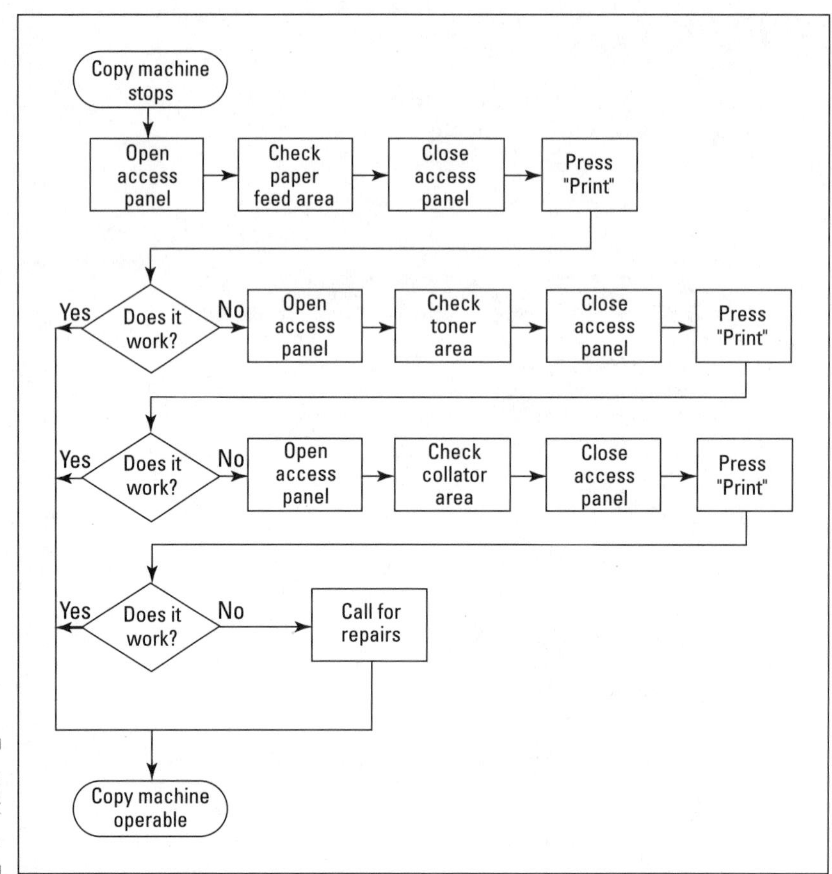

Example 5-19: A flowchart in action.

Table That Thought

Tables are columns and rows that display specific, related information. Tables carry more information per space than the same amount of text — yet they're often overlooked by technical writers. Find appropriate opportunities to create tables; they have great visual impact. Formal or informal, that is the question? Although there are no hard and fast rules about which tables should be formal or informal, use your judgment based on the formality of your document.

Fonts with personality

Each font has its unique personality, and you should let it shine. Use no more than two fonts, or your document will be too busy. Here are some tips on the use of fonts:

- Use a serif typeface (the ones you see here that have little feet) for text on paper documents. It's what we're used to seeing from the time we opened a first-grade primer. A popular serif typeface is Times Roman.
- When you generate electronic documents, a sans serif, such as Arial, is easier to read on the screen. Arial is also appropriate for headings on paper or electronic documents.
- You probably won't ever use funky fonts in technical writing, but who knows? If you have an appropriate use for such a font, you can choose from a wide selection of them that vary with the software. Play with them and save them for special occasions, such as announcing the company picnic.

On the formal side

Separate formal tables from the text with boxed headings, vertical and horizontal rules (lines), and a box, as you see in Example 5-20. If you use more than two or three tables in a document, assign a number to each. Place the table heading above the table. If you need to explain any information, place it below the table as a footnote.

When you think that the reader may have difficulty following a table across the rows, consider shading every other line, as you see in Example 5-20. You do that in Microsoft Word by choosing Format⇨Borders and Shading⇨Shading.

Table 4-5: Complexity Factors

Factor	Low	Moderate	High
Originality required		X	
Processing flexibility	X		
Span of operations	X		
Dynamics of requirements	X		X
Equipment		X	
Personnel	X		
Development costs			X
Processing time		X	
Communication architecture	X		

Example 5-20: Setting a formal table.

On the informal side

Informal tables are extensions of the text and don't have headings or table numbers. Merely write a sentence or two that has ties to the table. For example, in Example 5-21 you see a two-column table about risks that are inherent to a project.

Risks	Pinpointing discrepancies
Staffing	Staffing requirements and the staff available to fulfill those requirements.
Technical	Expected abilities of the technical platform and their actual abilities.
Scoping	Level of functionality and the time and resources available to develop the functionality.
External	Expected behavior of the environment outside the boundaries of the project and those inside the boundaries.

Example 5-21: Setting an informal table.

Check out Chapter 8 for a great way to create a procedure table when giving step-by-step instructions.

Go Figure

The difference between a table and a figure is simple. If a visual element isn't a table, it's a figure. Figures can be sketches (as you see in Example 5-22), drawings, photographs, charts, or graphs — in essence, anything other than columns and rows.

If you use more than two or three figures in a document, assign each a number. Include a concise title below or next to the text. Keep figures simple and uncluttered.

When you use a figure, make sure that it paints an accurate and clear picture. Example 5-23 shows a real-life figure that just boggles the mind. In this case, the thousand words would be better than the picture (to turn a phrase).

Example 5-22: The greater scheme of things.

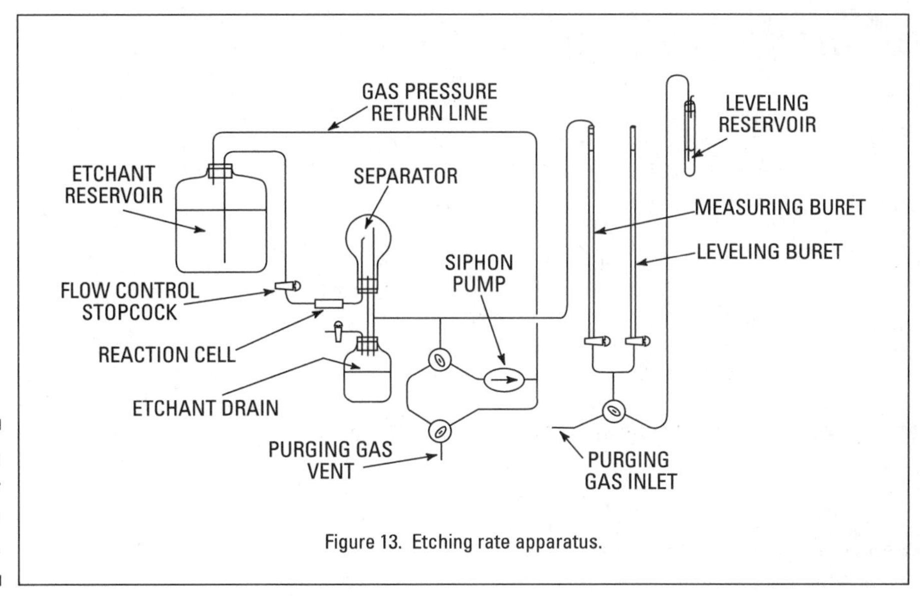

Figure 13. Etching rate apparatus.

Example 5-23: What's the point?

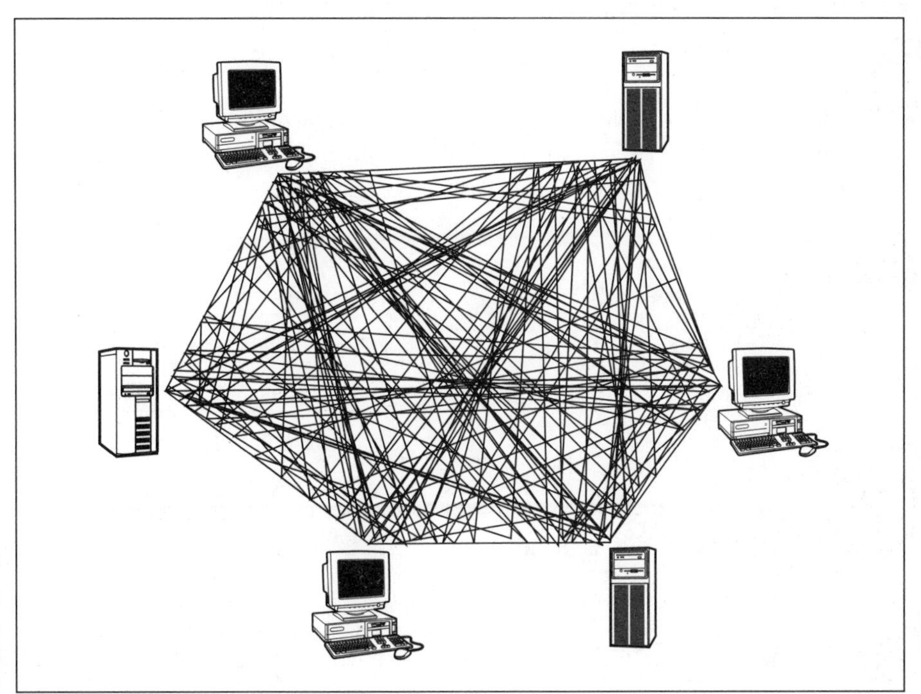

Why Is a Pink Slip Pink?

Colors evoke certain actions and reactions. For example, pink is known to have a calming effect on people. When an employer gives someone the ax, she wants the victim to remain calm and not do anything violent. Detention centers place prisoners in pink rooms to calm them down.

Color adds visual impact so that we can separate the ripe from the unripe, match our clothes, enjoy flowers, and so on. You can use color to create a mood and give a real-world look to the written word. With today's technology, you can invoke certain feelings by simply pressing a key. Press carefully! We've all been victimized by a shocking-pink Web site with gold letters. After viewing that for a few minutes, you need a headache remedy. Check out Table 5-2 to find out about the visual impact of colors.

If you use too many colors or use colors that are too bright, your documents may look more like circus posters. Pick one or two appropriate colors and stick with them.

Table 5-2 Colors and Their Visual Impact

Color	*Visual Impact*
White	Sanitary, pure, clean, honest
Black	Serious, heavy, death, elegant
Red	Stop, danger, excitement, heat
Dark blue	Calming, stable, trustworthy, mature
Light blue	Masculine, youthful, cool
Green	Growth, organic, go, positive
Gray	Neutral, cool, mature, integrity
Brown	Organic, wholesome, unpretentious
Yellow	Positive, cautious, emotional
Gold	Elegant, stable, rich, conservative
Orange	Emotional, organic, positive
Purple	Contemporary, youthful
Pink	Feminine, warm, youthful, calming

Meet Prints Charming

Many documents require photographs to get the point across. For example, photos are a great way to show people and equipment. (Photographs don't replace drawings when you want to show the internal workings of equipment.) You can scan a photo into your computer; crop distracting elements; prepare overlays by using arrows, letters, or numerals to point to certain elements; or retouch it to emphasize or delete certain details. Following are some photo tips:

- **Film:** If you plan to send the photo to a publication, find out what the publication prefers. Publications often opt for slides (color transparencies) and color negatives, rather than prints.
- **Black-and-white:** If you want to show sharp contrast, black-and-white may work better than color.
- **Caption orientation:** Position the caption with the same orientation as the photo so readers can read and see both without turning the page on its side.

Use relevant photos only. I recently saw a report written about superhighways. The publication displayed a full-page photo of a cement mixer that had nothing to do with the text. Perhaps the name on the cement mixer was the writer's cousin, and she was giving him free advertising.

Scale for Size

When it's important for readers to understand the size of an element, scale the element so readers can clearly envision it.

- **For non-technical readers:** Show something they relate to. In Example 5-24 you see a man's hand holding a small piece of equipment.
- **For technical readers:** Consider a measurement to scale the element. In Example 5-25, you see the element measured in microns. (A *micron* is one millionth of a meter.)

Example 5-24: There's no question of size.

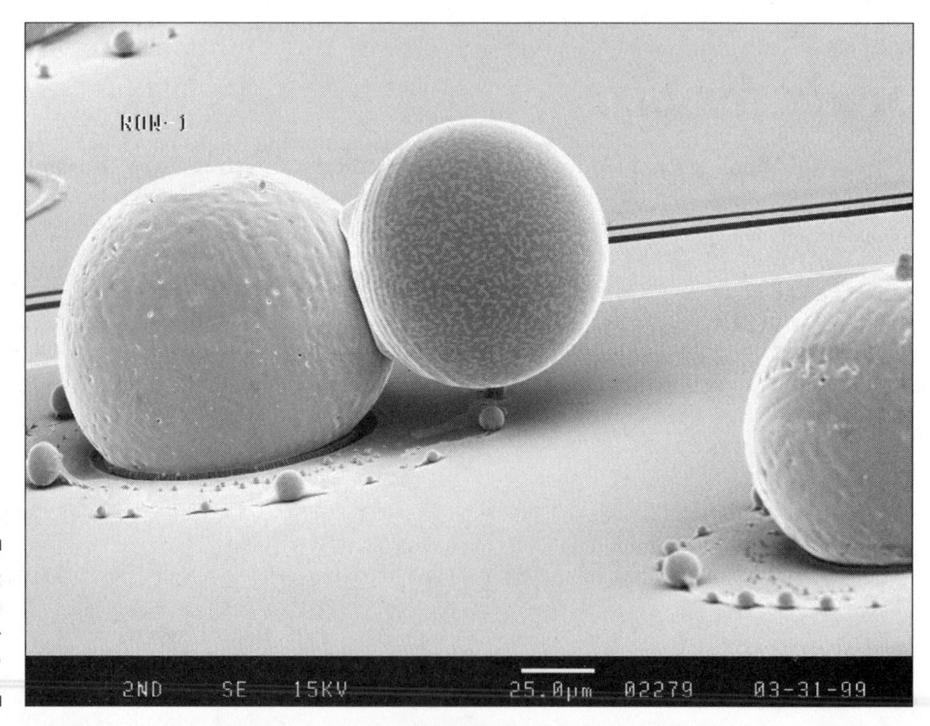

Example 5-25: Are these mountains or molehills?

Location, Location, Location

Place graphics — all graphics, not just photos — as close as possible to the related text. If you can't place the graphic on the same page as the text, try for a facing page or the next page. If this isn't possible and the graphic is lengthy, consider putting it in an appendix and cross-referencing it in the text.

What's Your Visual Preference?

The following examples demonstrate three ways to present the same data. They show the effect of a $10,000 investment with compound interest at 10 percent over 30 years.

- **Sentences:** Example 5-26 displays the data in sentence form. This format is useless because all the numbers are jumbled together and it's difficult to read.
- **Tables:** Example 5-27 lists the same information in table format. This format is a great way to display the data for someone who crunches numbers and needs the dollar amounts down to the penny (such as an accountant).
- **Bar charts:** Example 5-28 shows a bar chart for the "average Joe" who's interested in envisioning the growth over time.

Example 5-26: Get out the headache pain reliever.

The following information represents the growth of a $10,000 investment with compound interest at the rate of 10 percent over a 30-year period: Year 1, $11,046.49; year 2, $12,203.81; year 3, $13,481.66; year 4, $14,891.18; year 5, $16,452.76; year 6, $18,175.51; year 7, $20,078.64; year 8, $22,181.05; year 9, $24,503.60; year 10, $27,069.34; year 15, $44,536.55; year 20, $73,274.92; year 25, $120,557.49; year 30, $198,350.39.

Growth of a $10,000 Investment with Compounded Interest

Years Invested	Value of Investment
1	$11,046.49
2	$12,203.81
3	$13,481.66
4	$14,891.18
5	$16,452.76
6	$18,175.51
7	$20,078.64
8	$22,181.05
9	$24,503.60
10	$27,069.34

Example 5-27: Columns of growth down to the penny.

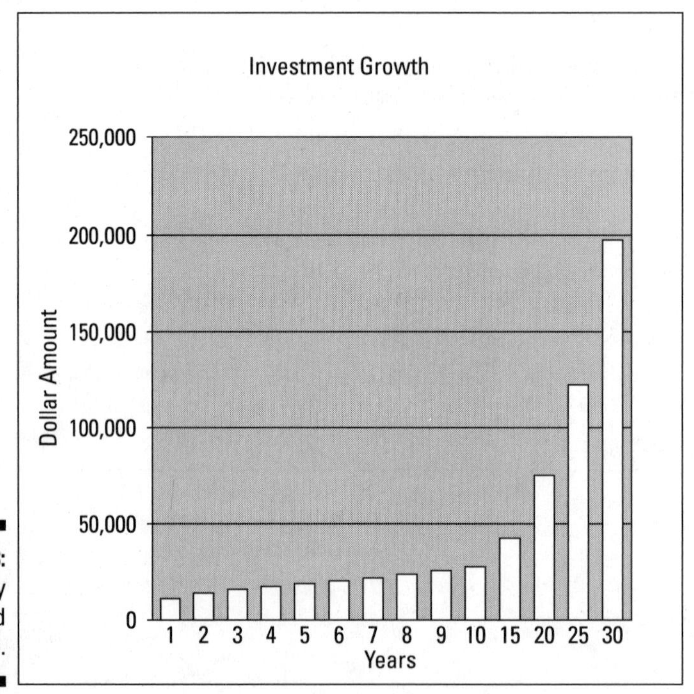

Example 5-28: Be wowed by the upward spike.

Chapter 6

Going In for a Tone Up

In This Chapter

- Keeping it short and simple
- Writing with a positive tone
- Using the active voice to bring life to your documents
- Choosing the right words
- Using humor and jargon appropriately

I watched his [Samuel Morse's] countenance closely, to see if he was not deranged . . . and I was reassured by other Senators after we left the room that they had no confidence in it [the telegraph machine].

—Oliver Hampton Smith (senator from Indiana, watching Morse demonstrate the telegraph to Congress), 1842

Readers aren't tone deaf. They "hear" the subtleties of everything you write. Say it simply. Say it positively. Say it actively. Say it consistently. Your goal is to help your readers use a product or understand a concept or process, not to dazzle them with embellished language. In essence, you must take technically mind-numbing concepts and convert them into something your readers understand.

Give 'Em a Little KISS

Keeping it simple is the epitome of honing the tone. KISS is an acronym that's used to represent **K**eep **I**t **S**imple, **S**tupid or **K**eep **I**t **S**hort and **S**imple. Although I prefer the latter, the gating word is *simple.* For example, instead of "olfactory impact," write *smell.*

The following "before" sentence is from an airline exit-seat card based on federal regulations. It's full of gobbledygook that doesn't add value. The "after" sentence is short and simple.

> **Before:** No air carrier may seat a person in a designated seat if it is likely that the person would be unable to perform one or more of the functions under REQUIREMENTS listed below. (33 words)
>
> **After:** To sit in an exit seat you must meet the following conditions. (12 words)

The essence of good writing is to be concise — shaving everything down to its bare essentials. Concise isn't the opposite of long; it's the opposite of wordy. If information doesn't add value, leave it out. Often, the less you say, the more impact it has.

KISS-ing technical documents

Clear and simple wording is important in technical documents because technical information — by its very nature — is difficult to read. If you surround technical words with simpler ones, your writing is easier to read. The following example illustrates that:

> **Concise:** Please confirm delivery of the HCE exchangers (5%/9%RD) needed by July 1, XXXX.
>
> **Wordy:** We wish to request that you notify us if the HCE exchangers (5%/9%RD) will be ready to be shipped and in our hands no later than July 1, XXXX.

Imagine that every word you write costs you $100. That gives you a motivation to cut to the quick. Every word that doesn't add to the effectiveness and clarity of your message just wastes the reader's time, reduces his energy, and costs you money.

Cutting to the quick

Why use several words when one will do? Check out Table 6-1 for ways to keep wording simple.

Table 6-1 KISS-ing Your Examples

Use	*Instead of*
Agree	Came to an agreement
Apply	Make an application
Breakthrough	New breakthrough
Conclude	Arrive at the conclusion
Consensus	General consensus
Consider	Give consideration to
Examine	Make an examination of
Experimented	Conducted an experiment
Investigated	Conducted an investigation
Invited (or asked)	Extended an invitation
Meet	Hold a meeting
Refer	Make reference to
Result	End result
Return	Arrange to return
Save	Realize a savings of
Show	Give an indication
Status	Current status

Using Contractions

Years ago, contractions belonged only in labor and delivery rooms — they were taboo in technical documents. Today, however, contractions are preferable because they add a personal, conversational tone to your writing. Also, if you write a negative, the apostrophe gives a visual clue. For example, *don't* is more obvious than *do not,* which may be mistaken for *do* at first glance.

Apply the conversational test to see whether a contraction works. Read your document aloud to hear how it sounds. For example, if you read the following sentence aloud, you would say *I am* instead of *I'm.*

I can't tell you how happy I am that the experiment was a success.

Don't use contractions in electronic documents because they may be difficult to read on low-resolution screens.

Accentuating the Positive

Is your glass half full or half empty? When it's half full, you're thought of as an optimist; when it's half empty, a pessimist. Let your readers know what they *can and will do,* not what they can't and won't do. Positive words engage the readers' goodwill and enhance your tone. The following sentences are positive and negative ways to send the same message. Notice the difference in tone.

Positive: We expect that you'll be *pleased* with the test results.

Negative: We hope you won't be *disappointed* with the test results.

The exception to this guideline is when you do want to accentuate the negative. For example, you may want to point a finger and say, "Theodore neglected to make one-fourth of the corrections," rather than, "Theodore made three-fourths of the corrections."

The glass is half full

The following words deliver a positive message. Look for opportunities to pepper documents with them.

Benefit	Bonus	Congratulations	Convenient
Delighted	Excellent	Friend	Generous
Glad	Guarantee	Health	Honest
Immediately	I will	Of course	Pleasant
Pleasure	Pleasing	Proven	Qualified
Right	Safe	Sale	Satisfactory
Save	Thank you	Vacation	Yes

The glass is half empty

The following words deliver a negative message. Avoid them when you can convey the same information in a positive way.

Apology	Broken	Cannot	Complaint
Damages	Delay	Difficult	Disappoint
Failure	Guilty	Impossible	Inconvenience
Loss	Mistake	Problem	Regret
Suspicion	Trouble	Unable	You claim
You neglected	Your inability	Your insinuation	Your refusal

Loving the Active Voice

Close your eyes and imagine this scenario: You're vacationing in the tropics with your loved one. The fiery crimson sun is slowly sinking into the distant horizon, and the waves are crashing over the craggy shore. You're sipping a glass of fine wine and affectionately clink your glass with your companion. That special someone leans over and whispers in your ear, "I love you." Don't those words create a warm, romantic atmosphere?

"I love you" is probably the most wonderful example of the active voice. It's animated and alive! What if that same special someone leans over and whispers in your ear, "You are loved"? (By whom? The dog?) Or worse yet, what if your special someone says, "You are loved by me"? (At that point, you'd probably start checking the personal ads.) The last two attempts at passion are passive. They're dull, weak, and absolutely ineffective.

Bringing life to your documents

Using the active voice is a major factor in projecting a tone that's alive and interesting. It's like looking the reader in the eye with authority and accountability. (*Voice,* by the way, is the grammatical term that refers to whether the subject of the sentence acts or receives the action.) In a sentence written in the active voice, the subject is the doer. In the following sentence, Dr. Salk is the doer — the discoverer. Can't you almost see Dr. Salk sitting in the lab totally absorbed in his work?

> **Active voice:** Dr. Salk discovered the polio vaccine.

When you write a sentence in the passive voice, the subject is acted upon. Sentences that use the passive voice are often dull and weak. In the following sentence, the vaccine is acted upon. Notice the difference in the impact between the two sentences. In the passive sentence, the actor is gone. This sentence doesn't conjure up much of a vision.

Passive voice: The polio vaccine was discovered by Dr. Salk.

Knowing when to use passive voice

Most of the time, when you use the passive voice, you come across as mealy-mouthed. There are times, however, when you may want to use passive voice because it's more appropriate. Do this for strategic reasons, not as a default. Following are appropriate uses of the passive voice:

- **You want to place the focus on the action, not the actor.**

  The shot was heard 'round the world. (The accent is on the shot, not the people who heard it.)

  The university was established in the early 1950s. (The accent is on the university, not those who established it.)

  Gloria was cited for her contribution to the outcome of the tests. (The accent is on Gloria, not the person who cited her.)

- **You're hiding something.**

  The tapes were erased. (This is a famous line that came out of the Watergate scandal that led to President Nixon's resignation. The passive voice was used so no one would take the rap for the 18½ minutes that were missing from the tape.)

Using Politically Correct Gender

It's savvy to be politically correct when you speak of men and women. For example, someone who's employed in a stockroom isn't a stockboy, but a stock clerk. Table 6-2 shows gender-neutral terms worth considering.

Table 6-2 Politically Correct Gender-Neutral Terms

Use This Term	*Rather Than This Term*
Ancestor	Forefather
Chair, moderator	Chairman, chairperson
Cinematographer	Cameraman
Delivery person, messenger	Delivery boy

Use This Term	*Rather Than This Term*
Firefighter	Fireman
Fisher	Fisherman
Flight attendant	Steward, stewardess
Humanity, human race	Mankind
Insurance agent	Insurance man
Letter carrier	Postman
Member of the clergy	Clergyman
Meteorologist	Weatherman
Nonprofessional	Layman
Police officer	Policeman, policewoman
Reporter, journalist	Newsman
Sales representative	Salesman, salesperson
Service technician	Repairman, repairwoman
Spokesperson	Spokesman
Synthetic	Man-made
Worker	Workman

When you speak of someone's job title, don't make a gender judgment. Judges aren't necessarily males, and nurses aren't necessarily females. If you can, identify the person by name. Moreover, a female adult isn't a girl — she's a woman.

Consider rewording the sentence

Gender neutrality is often a matter of rewording the sentence. The examples that follow show a number of ways to express the same thing:

Acceptable: Each person did the work quietly.

Acceptable: Each person worked quietly.

Acceptable: Everyone worked quietly.

Clunky: Each person did his or her work quietly.

Unacceptable: Each person did their work quietly. *(This sentence is grammatically incorrect.)*

Apologize in advance

When all else fails, you may consider using *he* (or *she*) to refer to both sexes. State your intentions (and apologies) at the outset of your writing. Another option is to alternate between the two. Notice how I address this delicate issue in the Introduction of this book. I use the male gender in the even chapters and the female gender in the odd chapters. (I didn't determine that women are odd. I flipped a coin.)

Consistency and Clarity Count

Following are guidelines for maintaining consistency throughout your documents:

- **Be consistent with wording.** For example, if you make reference to a *user manual,* don't later call it *reference manual, guide,* or *document.* Your readers won't know whether you're referring to the same publication or to different ones.
- **Don't replace technical terms with synonyms.** Repeating a word is better than compromising the integrity of what you write. For example, in the following "Compromised integrity" example, the writer changed "computer networks" to "computer systems." Although *system* is a synonym for *network,* the connotation is different. Therefore, the writer compromised the integrity of the sentence.

  Maintained integrity: The members of the networking group were learning all they could about computer networks.

  Compromised integrity: The members of the networking group were learning all they could about computer systems.

- **Avoid ambiguity.** For example, don't use the words *should* or *may* when there aren't options.

  Specific: Don't smoke when operating this equipment. (That's definite.)

  Not specific: You shouldn't smoke when operating this equipment. (There's a hint of "maybe.")

- **Be precise with words about locations such as top, bottom, left, or right.** Locations are subjective. In the examples that follow, the person may be looking over the computer from behind, so the switch would be on his left.

  Specific: As you face the front of the computer, you see the master switch on the right.

  Not specific: The master switch is on the right.

- **Use clockwise and counterclockwise to describe turns.** (Here's a little hint: It's a good idea to use symbols rather than words in technical writing because they're easy to recognize.)

  Specific: Rotate the dial 45° clockwise to create a seal.

  Not specific: Rotate the dial 45° to the right to create a seal.

Defining Terms

Use the language of your readers, and define any terms they may not understand. For example, in days of yore, a *mouse* was only "a small gray animal with a pointed snout, elongated body, and slender tail." Now, a *mouse* is also "a palm-sized device equipped with one or more buttons used to point at and select items on a computer display screen." Table 6-3 shows a variety of techniques you may use to help your readers understand technical terms. (Check out the Reader Profile in your Technical Brief to understand your readers.)

Table 6-3 Techniques for Describing Technical Terms

Techniques	*Example*	*Considerations*
Classical dictionary meaning	*Hydrocephalus* is defined as "an accumulation of serous fluid within the cranium, especially in infancy."	Be sure the reader understands all the terminology. For example, will your reader understand *cranium?*
Synonym	A "hummock" is a "hillock."	Unless the reader is familiar with either of these terms, he wouldn't know that you're talking about an *embankment.*
Antonym	*Antonym* is the opposite of "synonym."	If the reader isn't familiar with the terms *antonym* or *synonym,* your efforts are wasted.
Etymology	Have you ever wondered where the word *television* comes from? *Tele* is a Greek word meaning "distance," and *visio* is the Latin word for "sight."	Etymologies generally work well, especially for people well versed in languages.

Several years ago, I was trying to describe computing to my 80-year-old mother. She was having a difficult time understanding the difference between hardware and software. When I told her that "hardware is something you can kick," she got it immediately. Of course, this isn't a technical description, but it's one she related to. I understood my audience.

Who's Laughing?

Humor is a sensitive area, so use it cautiously. When handled properly, humor can make a technical subject more enjoyable and easier to understand. Here's how Lewis Thomas handled humor in *The Lives of a Cell:*

> "Ants are much like human beings as to be an embarrassment. They farm fungi, raise aphids as livestock, launch armies into wars, use chemical sprays to alarm and confuse enemies, capture slaves. The families of weaver ants engage in child labor, hold their larvae like shuttles to spin the thread that sews the leaves together for their fungus gardens. They exchange information ceaselessly. They do everything but watch television."

If there's the slightest chance that your humor may be misconstrued, avoid it. This is especially true when writing for people for whom English is a second language. What's humorous to you may be insulting to them.

When to Be a Jargon Junkie

Jargon is specialized shoptalk that's unique to people in an industry. Technical jargon is a hallmark of a good technical document to readers with a vast knowledge of the subject and the terminology. In these cases, watering down the language makes no sense. Doing so may damage the integrity of the document and insult the reader.

Be certain, however, that your readers (or the listeners in the following incident) know the language. I was on a flight waiting to take off when the pilot announced that takeoff would be delayed because of a problem with one of the lavatories. About 20 minutes later, he announced that all the trucks that could fix the problem (a pump out, I assumed) were busy. So he said we'd take off but shouldn't use the "aft lav on the port." Several people looked flustered. If you aren't an aviator or sailor and didn't read *Moby Dick,* perhaps you wouldn't know that *aft* means "rear" and *port* means "left."

Chapter 7

Dotting the *Eyes* and Crossing the *Tees*

In This Chapter

- Editing and proofreading for accuracy
- Understanding proofreaders' marks
- Using the editing checklist

Socrates was a famous Greek teacher who died from an overdose of wedlock.

—From a high school student's exam paper
(anonymous for a good reason)

If you don't think that a typographical error or faux pas can make a difference, take a look at the quote about Socrates. And what would you think if you went to a restaurant and saw this menu?

Full Coarse Meal
White Whine
Soap of the Day
Frayed Chicken with Raped Potatoes
-or-
Baked Zits
Tort of the House
Turkey Coffee

Now, imagine yourself spending days, weeks, or months writing your technical document only to have an error stick out like a tarantula on a piece of angel food cake. You'll be remembered for the error, not for the great document you spent days, weeks, or months writing.

Editing and proofreading are often-forgotten parts of the writing process, and the results can be devastating. In addition to giving people the wrong information, you will embarrass yourself.

Editing versus proofreading

There's a difference between editing and proofreading, although both result (hopefully) in a more readable and error-free document. Once you draft the document and the content is correct, it's ready for editing.

Editing refers to amending text by modifying words, sentences, paragraphs, or the general structure of the document. *Proofreading* is the final step before your document is ready for prime time. It refers to the systematic method of finding errors (such as typos) and noting them for subsequent correction.

Following are examples of editing and proofreading errors. Both appeared in Ann Landers columns.

Editing oversight: A bean supper will be held on Tuesday evening in the church hall. Music will follow.

Proofreading oversight: The rosebud on the altar this morning is to announce the birth of David Alan Smith, the sin of Rev. and Mrs. Julius Smith.

Here's a classic example of an embarrassing moment: A woman in one of my workshops was the director of public relations for a major corporation. She sent a quickly composed e-mail message to hundreds of people throughout the United States, Europe, and Asia. In her haste, this director of public relations left the *l* out of public. Can you even imagine her embarrassment?

Don't Turn On Your Computer and Turn Off Your Brain

Although your spelling and grammar checkers pick up a lot of errors, there's nothing like the human eye to see what is and isn't correct. For example, *due ewe sea any miss stakes inn this sentence?* Your spelling checker wouldn't.

Diamonds and faux pas are forever

Following are some tips for finding errors that your computer may not detect:

- **Double-check all names, including middle initials, titles, and company distinctions.** Many people get insulted when you misspell their names. Did you type *Lynn* when it should be *Lynne?* Did you write *Corp.* instead of *Co.?*

  Most names generally identify the sex of the person. However, you can't always make assumptions. For example, consider Stevie Nicks (female) and Fran Tarkenton (male). Who'da thought!

- **Double-check numbers.** Did you tell the reader the actual height is 7.43 inches instead of 74.3 inches? Did you refer to page 6 when the reference is on page 7?
- **Keep an eye out for misused or misspelled homophones (words that sound the same but are spelled differently).** Did you use *principal* when you mean *principle?*
- **Look for repeated words.** Perhaps you wrote, "I will call her her back in a week." Your word processor may not pick up the error; your e-mail application may not, either.
- **Be on the alert for small words you misspelled.** It's easy to type *of* instead of *if* and not notice the error. We tend to read what we expect to be there.
- **Check dates against those on the calendar.** If you wrote Monday, June 5, be certain that June 5 is a Monday.
- **Check for omissions.** Did you leave off a vital number or other piece of critical information?
- **Check spelling, grammar, and punctuation.** Use your eyes as well your computer tools. For more information about grammar and punctuation, see the appendixes.
- **Print out the message and reread the hard copy.** Why? We're all used to reading the printed word. Therefore, we tend to see errors on hard copy that we didn't notice on the computer screen. Also, with hard copy there's continuity.
- **Read the text aloud.** (Actually, mumble to yourself.) Can you read the document just once and thoroughly understand it? If you can't, reword what you didn't immediately understand.
- **Get a second opinion if the document is critical.** Ask an office buddy to take a look at it.
- **Read from bottom to top and/or from right to left.** Doing so lets you view each word as a separate entity and helps you find errors that you may otherwise miss.
- **Scan the document to see that it looks right.** Is the text aligned properly? Do the examples match the text? Are the graphics in the right place? Are numbered items sequential?
- **Place a ruler under lines of text to help you proofread lengthy material that you copy from paper to your computer.** It's easy to skip a line and never know the difference — especially when the text is statistical.

When you write for the Web, double-check the accuracy of all your links. Don't just generate the links; make sure they work.

Test your proofreading skills

So you think that you're a proofreading ace. Let's see. Read the following sentence *once* and count the *F*'s in the text. How many are there? You find the answer at the end of the chapter.

> FINISHED FILES ARE THE RE-
>
> SULT OF YEARS OF SCIENTIF-
>
> IC STUDY COMBINED WITH
>
> THE EXPERIENCE OF YEARS.

The Proof Is in the Proofreaders' Marks

When more than one person edits or proofreads a document, proofreaders' marks provide a standardized method of showing changes. (Proofreaders' marks convey changes when you review other people's documents, not your own.) Write these marks clearly and accurately so that anyone viewing the text understands your intentions. Example 7-1 shows the commonly used marks that you and anyone editing the document should use for easy recognition.

You may want to give a copy of the proofreaders' marks to anyone who reviews your document so you're all on the same page (so to speak).

Example 7-1: Commonly used proofreaders' marks.

Instruction	Mark in Margin	Mark on Proof	Corrected Type
Delete	℘	the ~~good~~ word	the word
Delete and close up space	℘	the wo~~r~~rd	the word
Insert indicated material	good	the ^ word	the good word
Let it stand	(stet)	the ~~good~~ word	the good word
Transpose	(tr)	the word good	the good word
Insert space	#	theword	the word
Close up	⁀	the wo rd	the good word
Period	⊙	This is the word	This is the word.
Comma	^	words words, words	words, words, words
Apostrophe	∨	Johns words	John's words
Double quotation marks	∨/∨	the word word	the word "word"
Uppercase	(uc)	the word	The Word
Lowercase	(lc)	The Word	the word

Use the Editing Checklist

Example 7-2 is an editing checklist to review before you finalize any document. It will save you many embarrassing moments! Remember the words of the ubiquitous Ann Onomyous: *The bitterness of poor quality remains longer after the sweetness of meeting the deadline has been forgotten.*

Editing Checklist

- ❒ **My headlines are compelling and will whet the reader's appetite.**
  - ❒ I've included a **key word**(s).
  - ❒ I'm telling a **story**.
- ❒ **The message has visual impact.**
  - ❒ Headlines are **informative**.
  - ❒ There is ample **white space**.
  - ❒ **Bulleted and numbered lists**, and **charts and tables**, were used where appropriate.
  - ❒ **Sentences** are limited to 25 words.
  - ❒ **Paragraphs** are limited to 8 lines.
- ❒ **I've reviewed the message for clarity, format, and style.**
  - ❒ The **message will be clear** to my reader.
  - ❒ The message is **logically organized**.
  - ❒ I've **sequenced** the message to keep my reader interested and moving forward.
  - ❒ The **tone** is clear and simple.
- ❒ **My spelling, grammar, and punctuation are correct.**
  - ❒ I used the **spelling checker**.
  - ❒ I checked my **grammar and punctuation**.

Example 7-2: Editing checklist.

Answer to Test your proofreading skills: If you counted **6**, you're right. Most people count 3, not paying attention to the word *of*, which appears 3 times.

Part III

Types of Technical Documents

"You're the tech writer interviewing for the Boxing Mayhem Game User's Manual? Here's your headgear and mouthpiece– Mr. Rosco's expecting you."

In this part . . .

Dearly beloved, we are gathered here to bury the typical, confusing, and ineffective way to write

- User manuals
- Abstracts
- Spec sheets
- Questionnaires
- Presentations
- Executive summaries

and to produce technical documents that impact your readers the way *you* want.

Chapter 8

The Ultimate User Manual

In This Chapter

- Getting started
- Understanding the needs of your readers
- Learning to chunk information
- Writing a procedure table
- Knowing what to include
- Getting feedback
- Understanding binding processes

What, sir, would make a ship sail against the wind and currents by lighting a bonfire under her deck? I pray you excuse me. I have no time to listen to such nonsense.

—Napoleon Bonaparte (speaking to Robert Fulton, inventor of the steamship)

The grim reality of writing user manuals is that no one really wants to read them. People refer to manuals when they have problems or need to figure out not-so-easy-to-understand functions.

Manuals can explain how to assemble, how to use, how to fix, and more. Manuals take a variety of forms. They can be in print or electronic media, or a combination of the two. Following are some of the user manuals you may come across:

- Tutorials that are self-study guides
- Training manuals used as textbooks
- Operator's manuals written for equipment operators
- Service manuals for technical repairs
- Maintenance manuals for semiskilled technicians
- Repair manuals for service technicians who handle extensive repairs

Get Up and Running

Before you start a project, hold a kick-off meeting for everyone participating in the process: subject matter experts (SMEs), writers, editors, production assistants, reviewers, and anyone else who will give or receive input. Check out Chapter 3 for more details on what you should do when you begin a project. Following are some of the issues to address at the kick-off meeting:

- **Objectives and scope of project:** Explain the goals of the project. Don't assume that participants know what they are.
- **Development methodology:** This involves the tasks and related activities. It also includes confidentiality issues, sign-off procedures, and audit trails.
- **Ground rules:** The ground rules cover everything from exchanging information to handling problems.
- **Roles and time frames:** Discuss people's responsibilities and how much time they have to complete their tasks.
- **Milestones:** Manuals often involve rounds of drafts and field testing. Include all the critical milestones. Plug in dates and the people who are responsible. Check out Chapter 3 for more information on documenting milestones.

Prepare a style guide when two or more writers are involved. All the writers need to use the same guidelines when it comes to style, formatting, fonts, placement of examples, and anything else that affects the layout and design. If the manual will be patterned after one that exists, make sure that every writer and reviewer has a copy. For an electronic manual, you can create Web page templates.

Assessing Your Reader

In order to understand your readers and their level of competency, fill out the Technical Brief you read about in Chapter 2 and see on the Cheat Sheet in front of this book. You must arrest any apprehensions your readers may have and provide instructions that are appropriate for their experience with the product or industry.

Animal, vegetable, or mineral?

The needs of your readers govern whether you deliver a paper-based manual, electronic manual, or combination. For example, Chapter 2 discusses details such as knowing the approximate ages of your readers. That's important because surveys show that older people are more comfortable with paper-based manuals. Table 8-1 lists the types of media and an example of what may be appropriate for each.

Table 8-1 Paper, Electronic, or a Combination

Medium	*Example*
Paper based	A paper-based manual is ideal for people who operate a piece of equipment in a factory. The workers don't have access to a computer and need to follow the manual to learn the equipment.
Electronic	People who are completely computer literate and can load a CD-ROM without instruction are the perfect audience for an electronic manual for a computer application. Once they load the CD-ROM, they find everything they need to know in electronic format.
Combination	Use a combination manual for a technical program where people need a little hand-holding to get started. For example, a paper manual helps them get started, and an online manual helps once they're working in the application. (Check out Chapter 18 for writing online help.)

Different strokes for different folks

Several years ago, I was asked by a major corporation to write a paper-based user manual. When I filled out the Technical Brief, I realized that 80 percent of the users were data entry folks, and 20 percent were high-level engineers — a ratio of 4:1. There was no way I was going to present a huge tome to data entry folks. They'd either faint or quit on the spot. To meet the needs of this diverse group of readers, I prepared the following as two separate manuals:

1. **Engineering User Manual:** I wrote the overall user manual for the engineers. It contained all the bits and bytes and nuts and bolts. This manual came to 500 pages, which I broke into major chapter headings, subheadings, sub-subheadings, and so forth. The company housed these svelte manuals in three-ring binders and needed to provide weight training belts for heavy lifting.

2. **Data Entry User Manual:** I wrote a separate 40-page manual for the data entry people. It included all the tasks they needed to perform and a step-by-step approach to performing them. Examples 8-1 and 8-2 show a two-page spread from that manual. Example 8-1 discusses the parameters. Example 8-2 gives step-by-step instructions for the task.

   I had this manual bound separately (with its own cover) and three-hole punched so it could be inserted into every engineering manual. Because the ratio was 4:1, I had 4 data entry manuals printed for every engineering manual so data entry folks could have their own copies. I also prepared a Reference Card for the data entry folks.

The company for whom I wrote these manuals had them evaluated by professional reviewers. Both manuals got rave reviews for presentation and ease of use. The company also got great feedback from the engineers and data entry folks.

The Devil Is in the Details

You must write instructions with clarity and keen attention to detail. Never assume your reader will read between the lines or read your mind. For example, the following figure shows nine dots. Join them together with four straight lines without lifting your pen(cil) from the paper. Please try it.

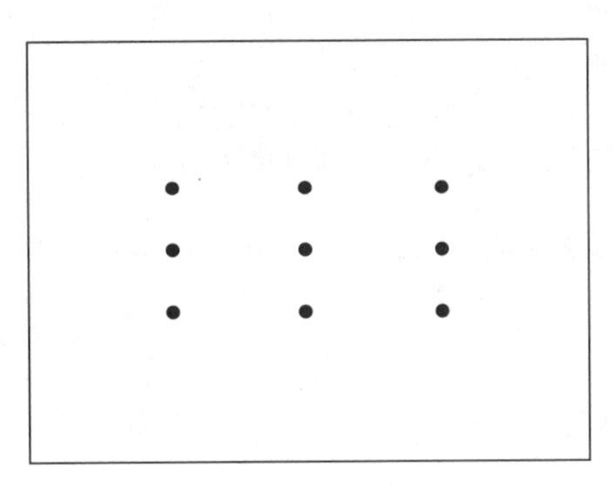

Did you do it on the first try? If not, that's because I wrote vague instructions. When you write vague instructions, your reader can't complete the task successfully. Look at the end of this chapter for how to connect the nine dots with four lines.

Check Out the Contents of the Box

Here's a scenario you may relate to: You buy a new gas grill at a greatly reduced pre-season price. You can't wait get home to smell the aroma of your juicy red steaks sizzling on the grill. When you open the box, however, you notice that the grill is completely disassembled. The instructions say, "This is a do-it-yourself project that any 5-year-old can put together." It looks as if the sizzling will have to wait until you assemble the grill.

You spread out the grill parts on the ground but can't tell whether you have all the pieces. The instructions don't tell you what parts are enclosed. You won't know whether you have the right parts until the grill doesn't work or you have parts left over.

The lesson learned is this: Always tell your reader what's in the box (or packaging) so there are no surprises. It's a good idea to list all the parts and show photos or drawings of any parts your reader may not recognize.

SELECTING THE LINKEASE FAX[SM] SERVICE (DOMESTIC AND INTERNATIONAL)

Introduction You can send a domestic or international FAX message to any Group III facsimile machine. This is a wonderful convenience if you need to send a FAX to anyone who doesn't have LinkEase or a telex terminal. You have the option of sending a FAX cover sheet and can accommodate FAX machines with various paper sizes.

How to Address a Domestic and International FAX

FAX Service	Description
Domestic	A telephone number. **Example:** 207-555-1234 *Note:* There is no need to put the number 1 in front of the telephone number, but *you must include* the area code — even if it is the same as yours.
International	A telephone number, starting with the numbers 011, followed by the country code, city code, and telephone number. **Example:** 011-4122123456 *Note:* In the example above, 011 is the required FAX code, 41 is the code for Switzerland, 22 is the telephone number of the FAX machine.

Message Parameters

- The maximum character count is 300,000 (12 typed, single-spaced pages).
- The page length can be 11 inches or 14 inches.
- The maximum line width is 90 characters or 132 characters.

Delivery FAX delivery is available to any Group III FAX machine in the United States and over 150 foreign countries. Most LinkEase FAX messages are delivered within minutes of their receipt by LinkEase. LinkEase will make several attempts to deliver your message. If the FAX machine is not reachable (perhaps it is busy or out of order), LinkEase will cancel the message and notify you with a message to your Mailbox.

If you address your message to a Group I or II FAX machine, LinkEase will make every effort to deliver your message; however, a surcharge will apply because of the slower reception/transmission times of these machines.

Example 8-1: All you need to know about the task.

SELECTING THE LINKEASE FAX SM SERVICE (DOMESTIC AND INTERNATIONAL) (continued)

Procedure The following procedure tells you how to send a LinkEase FAX.

Step	Action
1	Follow steps 1-6 in the section entitled Preparing the Address.
2	Press the [down arrow] until FAX is highlighted and press [RETURN]. *Result:* You'll see the FAX screen.
3	Type the FAX number and press [RETURN]. *Note:* If you're sending an International FAX and don't know the country code, press [F2] and you'll see the Country Code screen. Type in the country, press [RETURN], and the country code will be added to the FAX number.
4	Press [RETURN] for a standard 90-character, 11-inch length. *Options:* • Type **L** + [RETURN] for a 90-character, 14-inch length. • Type **W** + [RETURN] for a 132-character, 11-inch length. • Type **WL** + [RETURN] for a 132-character, 14-inch length.
5	Type the attention line, if applicable, and press [RETURN].
6	Do you want to Add to the Dictionary? • If *yes*, press the letter **Y.** • If *no*, press the letter **N.** *Note:* If you want to add to the Address Directory, see section 4.6 in the Reference Manual.
7	Press [Tab] to continue. *Result:* A Message File Selection menu will appear.
8	Follow the steps in the following sections: • Creating the Text. • Sending the Message.

Example 8-2: A step-by-step approach.

Chunking Up

After you assess your readers, think of the tasks they need to perform. Then break the user manual into manageable, bite-sized chunks of information. For example, in a word processing application, readers need to perform the following tasks:

- Compose, save, edit, and revise
- Cut, copy, and paste
- Print the document, a page, or a selection
- Understand the menu commands

Once you identify all the tasks, break them into separate chapters and subchapters. Example 8-3 shows the outline for a chapter with associated tasks for maintaining an address list in an e-mail application.

Chapter 5: Maintaining an Address List

Address Lists
Creating a New Address List
Required Information by Message Type
Listing All the Entries
Searching for a Name or Group
Updating an Address List
Sending a Message to More Than One Address
Sending and Receiving Address Lists
Using an Address List as a Database File

Example 8-3: Outline of a sample chapter.

The long and short of it

In my writing workshops, I'm frequently asked how long should a user manual be. My stock answer is: *Say what you need to say and get out.*

For example, I recently purchased a backup tape drive for my computer. It came with a wonderful manual printed on a single sheet of 11-x-17-inch heavy card stock. The paper was printed on two sides and folded in fourths. This brief manual walked me through the steps for installing, connecting, backing up, and troubleshooting. It was easy to read and told me all I needed to know.

Table That Procedure

When you write a procedure, you're the teacher — the expert who needs to share your knowledge in a logical, step-by-step format. Check out Example 8-2 to see how effectively this can be done.

If you opt to use a different format for the table, create a standardized method for your readers to identify procedures at a glance. Assign each step a number and start with an action word — something readers should do.

Correct:

1. Place the cursor in the field in the first line.
2. Press [Shift] and [F5] at the same time.

   Result: The special handling attribute will be assigned automatically.

Incorrect:

1. Place the cursor in the field in the first line.
2. Press [Shift] and [F5] at the same time.
3. The special handling attribute will be assigned automatically. *(This isn't a step; it's the result of a step.)*

Between the Covers

When you prepare a lengthy user manual, include all the information readers need and make the information easy to find. Following are guidelines for what to include.

Table of contents

Whenever the manual is more than 15 pages, include a table of contents to help readers easily find what they need. Use leaders (. . . . . . .) to connect the subjects to the page numbers so readers can run their eyes across the page.

Some writers use internal chapter numbers. For example, Chapter 1 is numbered 1-1, 1-2, 1-3, 1-4; Chapter 2 is numbered 2-1, 2-2, 2-3, 2-4, and so forth. This is different from consecutive numbering (1, 2, 3, 4). Use internal chapter numbers when you anticipate updating one or more chapters. Doing so keeps you from renumbering (and reprinting) the entire manual.

Glossary

If there's a hint that your readers may not understand all the terminology in the manual, include a glossary at the end. Defining a word the first time you use it isn't always adequate because readers won't read your manual from cover to cover as if it's the great American novel. They'll use it for reference when they have a question.

If you write online help, use pop-up windows for terms. Check out Chapter 18 for details. If you write Web documents, create a link to the glossary.

Index

Not all manuals need an index. The content of the manual is the gating factor, not the length. Put yourself in the mindset of the reader and ask: If I were reading this document, would an index be helpful? If you think an index would help you, include one. *Be very sensitive to the logical search words that the reader may look for.*

I heard a story of a woman who bought a new car. She drove about 25 miles when she ran over some glass and got a flat tire. The woman wanted to change the tire herself and pulled out the owner's manual. She checked under the following letters in the index:

f for flat

t for tire

j for jack

c for change

The woman couldn't find the information and couldn't believe that the manual neglected such an important entry. She finally gave up and phoned the American Automobile Association (AAA). When she returned home, she began checking every index entry. Yes, there was an entry for changing a flat tire. It was under *h* for "How to change . . ."

When writing for the Web, include a site map or something equivalent.

Troubleshooting

No matter how well you write the manual, readers will have issues. The way to address these issues is to include a section on troubleshooting. Example 8-4 is a great troubleshooting section from the *Kodak DC215 Zoom Digital Camera User Guide.* It breaks out the problem, cause, and solution.

Camera		
Problem	**Cause**	**Solution**
Picture is too light.	The flash is not needed.	Change to Auto flash. See page 13.
	The subject was too close when the flash was used.	Move so there is at least 1.6 ft (0.5m) between the camera and the subject.
	The light sensor is covered.	Hold the camera so your hands or other objects do not cover the light sensor.
	The Exposure Compensation is set incorrectly.	Decrease the Exposure Compensation. See page 14.
Stored pictures are damaged.	The camera memory card was removed when the Ready light was blinking.	Make sure the Ready light is not blinking before removing the card.
Pictures remaining number does not decrease after taking a picture.	Image Resolution and Quality settings do not take up sufficient space to decrease the picture remaining number.	The camera is operating normally. Continue taking pictures.
Picture is not clear.	The lens is dirty.	Clean the lens. See page 65.
	Subject was too close when picture was taken.	Stand at least 1.6 ft (0.5m) in wide angle, 3.3 ft (1m) in telephoto.
	Subject or the camera moved while the picture was taken.	Hold camera steady until the picture is taken.
	The subject is too far away for the flash to be effective.	Move so the subject is less than 10 ft (.3m) away.

Example 8-4: Getting rid of troubles.

Making a list and checking it twice

After you write the document and *before* you send it for testing, check it for basic proofreading and editing errors (see Chapter 7). Then answer the following questions:

- Is the manual complete?
- Is it well organized?
- Is it technically accurate?
- Did I explain terms or abbreviations that readers may not understand?
- Did I use consistent language?
- Are the examples in sync with the text?
- Are the examples placed where they belong?
- Do the subjects and page numbers in the text match the table of contents and index?

When you include a section on troubleshooting — or frequently asked questions (FAQs) in an electronic document — you spare the folks at the Help desk cauliflower ear. Here's how to think through what to include in the troubleshooting section:

- If this is the first edition of the manual, think of the questions or problems your readers may have. Be sure to include questions that testers had during alpha and beta testing.
- If this is a second or third edition, ask the people at the help desk for the most commonly asked questions.

Feedback form

To improve the quality of your manual for the next printing and to improve the quality of your own writing, include a feedback form. This form gives readers a chance to let you know what they liked and didn't like about the manual and to let you know of any errors they found. Check out Chapter 11 for tips on writing a feedback form, which is a type of questionnaire.

Don't be disturbed if you get mainly negative feedback. Readers who are pleased with the manual generally don't bother to respond. Readers who aren't pleased want to let you know. Take this feedback as constructive and use it as an opportunity to improve your writing technique. Don't take it personally.

Testing, Testing, 1-2-3

You won't know whether your manual is accurate and understandable until you have it tested extensively. Ask your testers to keep a detailed log of everything that doesn't work, is unclear, or is wrong. Following are a number of approaches:

- Get a subject matter expert (SME) to review the manual for technical accuracy.
- Test the manual in typical user environments, if possible. For example, if you write a manual for Windows, Mac, and Linux environments, have people test in those environments.
- Go with a professional testing group such as National Software Testing Labs (NSTL), `www.nstl.com`; SysTest Labs, `www.systest.com`; or VeriTest, `www.veritest.com`.
- Ask a novice user to read the manual to make sure that it's accurate and easily understood. A novice often finds a need for explanations or instructions that you take for granted.

Years ago, I wrote a user manual for a company developing accounting software, and I asked a novice to test what I wrote. When he tried to press a numeric key, his fingers headed straight for the numerals on top of the alpha keyboard. He couldn't get the numbers to show up on the computer screen. The problem? I was accustomed to using the keypad and neglected to mention that the user needed to use the keypad, not the "typewriter" numeric keys. The moral of the story is to take nothing for granted.

Getting All Bound Up

Although most people opt for a loose-leaf binder, you have a number of binding options for your manual. Check out Table 8-2 for a variety of binding options and when they're appropriate:

Table 8-2 Binding Options

Binding Option	*How to Use It Effectively*
Loose-leaf binder	A two- or three-ring binder is appropriate for documents that are 100 pages or more or documents that need to be updated more than once a year. It's easy to slip pages in and out.
Perfect binding	Perfect binding means that the pages are glued together much like a paperback book. If you use this method of binding, use left and right margins that are 1½ inches because the book is trimmed after it's glued together. The disadvantage to perfect binding is that you can't keep the manual open to a certain page without weighting it down and eventually damaging the binding.
Saddle stitching	Saddle stitching is another name for stapled in the spine. Saddle stitching gets its name because the stapler looks like a saddle (^) in the center. You place the pages on the ^ and staple away. This is a good and attractive option for short manuals of 100 pages or less.
Spiral binding	Spirals come in plastic or wire and are an inexpensive way to bind manuals of fewer than 100 pages. The advantage of spiral binding over saddle stitching is that the manual stays open to any page without cracking the spine.

Join-the-Dots Brain Teaser

The instructions in "The Devil Is in the Details" earlier in this chapter were vague. If they were detailed, one of them would have told you where to start and another would have told you that you don't need to stay within the parameters of the square. Here's the solution:

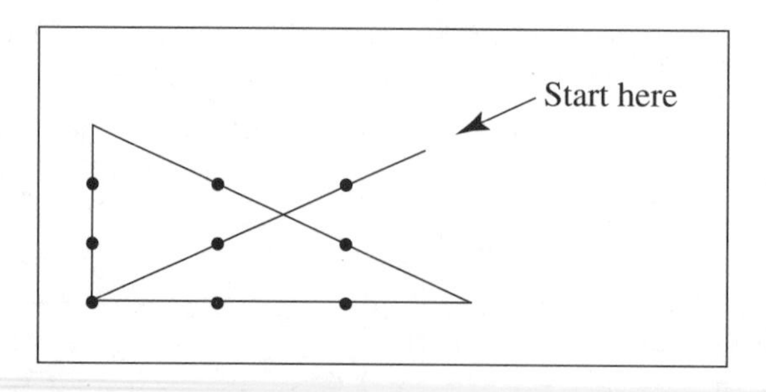

Chapter 9

Writing in the Abstract

In This Chapter

- Understanding the types of abstracts
- Using abstracts appropriately

The factory of the future will have only two employees, a man and a dog. The man will be there to feed the dog. The dog will be there to keep the man from touching the equipment.

—Warren G. Bennis, American educator and business writer

You write an abstract to condense the key issues of a longer document. It's somewhat like a preview intended to whet your readers' appetites. This is similar to the trailer of a movie that helps viewers determine whether they want to see the full feature. You don't write an abstract for all your documents, only for documents that are very long or highly technical.

Because busy readers use abstracts to determine whether they want to read the entire document, well-written abstracts give readers a brief, high-level description of the topic. Abstracts become critical segments of articles or reports.

Types of Abstracts

Abstracts fall into two categories, depending on the type of information they convey. An abstract is either *descriptive* or *informative*.

Descriptive abstracts

Descriptive abstracts are generally very short writings of a few sentences. They're informal and may merely be a table of contents in sentence format. And they don't have a headline that says "Abstract." The following descriptive abstract is taken from a user manual and shows how informal a descriptive abstract may be. It tells you in just two sentences what the user manual is about.

> *This user manual describes the commands, statements, functions, and uses of Wonderword word processing software. This software runs on Windows 97 or higher.*

Informative abstracts

Informative abstracts provide the readers with key aspects of the document. You write informative abstracts to stand alone or to be part of a lengthy paper. Limit abstracts to between 200 and 250 words or less.

Example 9-1 is the first page of a paper that appeared in a technical publication prepared for the IMAPS 2000 Proceedings, held September 20–22, 2000, in Boston, Massachusetts. The brief abstract introduces the paper in the opening paragraph that spans both columns. The entire article was bound in a publication that was distributed at the conference. The author presented the paper at the IMAPS conference. (To learn more about presenting a paper, check out Chapter 12.)

What to include

Following is what to include in an informative abstract:

- Subject, scope, and purpose of the study
- Methods used
- Results
- Recommendations, if any

What to omit

Following is what you omit from the abstract. If the reader needs this level of detail, she'll read the full text.

- A detailed discussion of the methods
- Illustrations, charts, tables, or bibliographical references
- Any information that doesn't appear in the full text

Abstract portion

Stencil Printing Holds High Promise for Wafer Bumping

Jon Roberts
Cookson Performance Solutions
Foxborough, MA 02035
Phone: 508-698-7225
E-mail: jroberts@cps.cookson.com

Abstract

Solder bumping for semiconductor wafer applications requires scaling a stencil printing process from the current 50-mil geometries downwards by an order of magnitude, while driving defect densities even lower to maintain high yield. In particular, wafer bumping moves the process to smaller area ratios, where the print covers a smaller area on the wafer, but using a thicker stencil to achieve a high print volume. The effects of aperture periphery begin to dominate the printing quality, and the paste particle size approaches the of the stencil thickness. A successful process calls for an integration of printing equipment technology, solder paste development, stencil manufacture improvements and reflow furnace advances. In this paper, we describe some of the metrics used to evaluate these components and give early results of some of the tests. This work helps to determine directions for further refinement.

As a packaging choice for IC's that's been around for decades, flip-chip with solder bumps has witnessed improvements in many of the alternative die assembly processes. Although solder bump users have made their own advances, the benefits just haven't yet outweighed the problems and costs involved with its implementation. Electroplating (Solders or gold) and evaporation/sputtering as methods of making bumps haven't kept up with the advances in automated wire bonding.

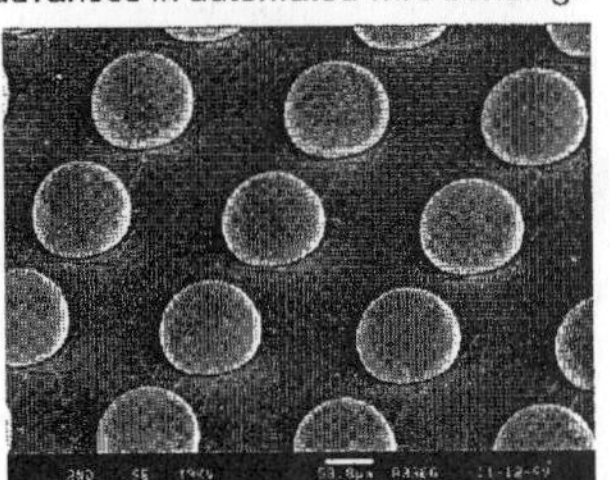

Figure 1 Typical printed solder bumps after reflow and cleaning

Stencil printing, however, brings promise of high yield and throughput, low tooling costs, and full automation to the competition. Including wafer handling and printing, a production rate in excess of 40 wafers per hour is easily achieved. Adaptable for a variety of solder paste compositions, there is no penalty for wafer size evolution and printing speed is independent of pattern density and bump size. Figure 1 shows eutectic solder bumps printed with Alpha Metals's WS 3060 paste to a pattern at 10-mil pitch, reflowed, and cleaned, ready for flip-chip or direct chip attachment.

Process integration for wafer printing calls for more than just scaling down the stencil dimensions. The typical aperture size, about 125 microns, violates the aspect ratio rule for stencils thicker than 85 microns. But high-density patterns to make large bumps require large, closely spaced apertures in thick foils. Getting past the aspect ratio rule puts increased demands on the paste release and stencil wipe processes and frequency. Reflowing the paste into uniform bumps requires the collection of all printed paste beads into the melt – flux chemistry and reflow atmosphere control are critical to successful reflow. And reflow must leave flux residues which can be cleaned with chemistries friendly to semiconductor wafers as well as to the environment.

Stencil Development is the Key to High Yields

The stencil represents a critical limiting factor in the quality of the printed wafer. The reflowed bump's size variation can be no better than the variation of the aperture size. Although the reflow process will re-locate a paste brick to perfect position on the UBM pad, the alignment of the aperture to the pad

33rd International Symposium of Microelectronics IMAPS 2000 Proceedings
September 20-22, 2000 Boston, Massachusetts

Example 9-1: Highlights in abstract form.

Writing in style

In the opening sentence of the abstract, announce the subject and scope. Then you must decide what's relevant and follow with the major and minor points. Write clearly and concisely. Check out Chapter 6 to hone the tone and keep the text short and simple.

Don't forget to give yourself a byline — your name under the title. Below the title, type your name and the names of anyone else who co-authored the abstract. You may list coauthors alphabetically or in the order that represents their contributions.

Using Abstracts Effectively

Abstracts can stand alone or be part of the longer text. When they stand alone, always let readers know where they can find the full document — in print and/or electronic form. Following are examples of how abstracts may be used:

- Some companies distribute abstracts at trade shows, conferences, or seminars to generate interest in their products or services.
- Abstracts are often sent to management-level folks (inside or outside the company) to give them a thumbnail version of a topic.
- Abstracts appear in journals such as *Chemical Abstracts* with information on how to find the actual document. (You get an abstract published in a journal in much the same way that you get an article published. Check out Chapter 20.)

When an abstract is part of the longer text that appears in a professional journal, it precedes the article. When the abstract is part of a report, place it immediately following the title page.

There's a big difference between an abstract and an executive summary. An executive summary covers the information in greater detail and may include charts, graphs, or other visual aids that summarize the full text into a few pages. An executive summary never stands alone; it's always part of the body of a long report. Check out Chapter 13 for more information on writing an executive summary.

Chapter 10

Writing Spec Sheets

In This Chapter

- Requirement specs
- Functional specs
- Design specs
- Test specs
- End-user specs

It will only be a matter of time before humans are tattooed with a similar mark to the codes in the supermarket.

—Cary H. Kah, *En Route to Global Occupation (1991)*

Spec sheets (the shortened version of specification sheets) cover a wide variety of products and equipment, including software. Spec sheets may be written by engineers, technicians, programmers, or other technical people. If these folks don't have the know-how or the time, the technical writer may be asked to assist with the writing process. Everyone involved in developing a project relies on spec sheets for all the details of the project, much like everyone involved in constructing a building relies on the blueprints.

As you write technical documents (such as user manuals, online help, and more), use the spec sheets to answer questions you may have. If the spec sheets don't answer your questions, your next line of defense is the subject matter expert (SME).

The Natural Order of Things

You write spec sheets in sequential order because one builds on the other. Spec sheets are always works-in-progress and need to be updated as the project changes. (To relate this to building a house, the architect may prepare a rough sketch that doesn't include the plumbing, heating, and other details. Then the architect prepares the blueprints. He changes the blueprints each time the homeowner changes the design.)

Following is the order in which spec sheets may be developed:

Phase 1: Requirement specs

Phase 2: Functional specs

Phase 3: Design Specs

Phase 4: Test Specs

Phase 5: End-User Specs

When you write spec sheets, you work in collaboration with technical people who help make the decisions about what to included. Therefore, the spec sheets in this chapter are guidelines; they're not carved in granite.

Phase 1: Requirement Specs

When a company plans to introduce or update a product, it writes requirement specs to provide the design group with a starting point. These specs may be an outgrowth of a marketing analysis that mirror the needs of the marketplace. At the very least, requirement specs include the following information:

- **Definition of the application or product:** Detail all available information about the application or product.
- **List of functions and capabilities:** Include information about what the application or product is capable of doing.
- **Estimated cost of finished product:** Create a ballpark estimate of what the application or product should cost to make it competitive in the marketplace.

Phase 2: Functional Specs

Functional specs expand on the list of capabilities iterated in the requirement specs. They deal with how the system operates. Here's some information to include in a functional spec:

- **System overview:** Explain the objectives, capabilities, methods, system operation, output, and what the system will and won't do.
- **Data dictionary:** Include all the facts and figures that enter the system or are produced by it. This includes data names, abbreviations, data sources, range for numerical data, definitions or data descriptions, and codes and their meaning.
- **Input description:** Describe data that comes into the system and how the end user enters the data. This is key to the spec because the users are responsible for input; therefore, everything is geared toward ease of use.
- **Operation description:** Explain what the system will do, under what conditions, and what will result.
- **Calculations:** Include formulas to determine how the system generates numbers for output.
- **File description:** Describe the purpose, content, use, and structure of the data files.
- **Other stuff:** Explain anything that's unique to the application or product.

Phase 3: Design Specs

If you think functional specs are long, wait until you see the design specs. They add yet another level of detail. Design specs may include any or all of the following:

- **Relevant documents:** Include documents that have information relevant to the application or product. This is a critical resource to the technical writer who writes the manual.
- **Functional description:** Include a detailed description of the functionality. This description may be in the form of words or diagrams.
- **Interfaces:** Discuss interfaces to the product, including power requirements and more.

- **Programming considerations:** For software, deal with all aspects of the functionality the programmer works with.
- **Reliability:** Describe how reliable the product is and how often it should be serviced, maintained, or updated.
- **Diagnostic issues:** Describe the testing and evaluating required to ensure a quality product.
- **Deviations:** Describe changes that may be necessary as the project progresses.

Phase 4: Test Specs

Before you bring hardware or software to market, you must test the product under various conditions. Therefore, when you write test specs, you should include the following:

- **Relevant documents:** Include any documents that relate to similar hardware or software you've already developed. These documents may be helpful to testers.
- **Product description:** Identify what is and isn't being tested.
- **Testing method:** Provide a step-by-step description of the testing procedure. This must also include the process for recording and reporting problems.
- **Precautions:** Identify any special care or issues the tester must deal with. For example, you may want to document a secondary procedure to accomplish something (called a *work-around*) in case the primary procedure doesn't work.

Phase 5: End-User Specs

End-user specs are basically product information sheets that ship with the product. They give users information about running the software or operating the equipment. Example 10-1 shows end-user specs that may accompany a software application. For other applications, end-user spec sheets may include features, strengths, weaknesses, product characteristics, support vendors, and more.

I. System Requirements

- Platform: MS Windows 95/98/00, or NT
- Memory: (95/98) 64MB; (NT) 128 MB; once installed in memory, KnowledgeSync takes approximately 17 MB of memory
- Processor: Pentium 200 (or greater); either platform
- Disk Space: (For installation) 45 MB maximum if no Windows ".dlls" are installed
- Disk Space: (For the application) 16 MB
- Disk Space: (Per event) 5k maximum (this includes event, query, subscribers, and all message texts)

II. Programming Languages

- C++, Visual Basic
- Microsoft COM (for API)

III. Frequency of Bug Fixes and Minor Enhancements

- Non-Critical: Every two months
- Critical: As needed

IV. Frequency of Major Releases

- Two times per year

V. Maintenance Program

- 20% of product license price
- Also covers version upgrades

Example 10-1: Specs for the end user.

Chapter 11

How Am I Doing? That Is the Question(naire)

In This Chapter

- Understanding the types of questionnaires and their advantages
- Designing a questionnaire
- Choosing the types of questions to ask
- Benefiting from the responses

> *The census is taken by the Census Bureau every 10 years, as required by the Constitution. (For the other nine years, Census Bureau employees play pinochle while remaining on red alert, in case the Constitution suddenly changes.)*
>
> —Dave Barry, "Counting the House," *The Boston Globe,* 4/9/00

The U.S. Census is the mother of all questionnaires. It finds its way into the mailboxes of American households every ten years. The census form may quiz us about everything from our ethnic backgrounds to whether we wear thong underwear. (Well, maybe it doesn't get quite that personal, but the long form isn't far from it.)

In this chapter, you learn to design questionnaires so you get feedback that's meaningful. What follows are situations for which you may use a questionnaire (also called an *evaluation form* or *survey*):

- Prove or disprove a theory
- Determine the success of a workshop, seminar, or presentation
- Measure job satisfaction

- Understand if you met your readers' needs (as in a user manual)
- Anticipate the success of a new product or service
- Form conclusions about a host of other issues

Don't Shoot the Messenger

Whatever "messenger" you use to deliver your questionnaire (face-to-face, mail, or the Web) design your form to get as much information as possible by making it easy for respondents to reply.

Getting up close and personal

Face-to-face questionnaires give the interviewer an opportunity to follow up on answers and delve more deeply into details. This is a great way to speak with people personally to ask what they think of an idea or product. This type of questioning, however, is limited to the number of people the interviewer can reach, so the results don't always represent a large cross section of people.

Some people are put off by face-to-face questionnaires because they sit there twiddling their thumbs while the interviewer jots down every word. We've all been approached by people in shopping malls who want to "take just a few minutes" of our time. Three hours later we're still held captive.

Using mail or the Web

There are advantages to questionnaires that are mailed or appear on the Web. A key advantage is that the interviewer can't influence answers by facial expressions or tone of voice. Therefore, the results tend to be unbiased. One key disadvantage is that people with strong opinions on a subject are more likely to respond, so results may be skewed. Another drawback is that there's no opportunity for someone to follow up with pertinent questions that may shed more light on a subject.

Be aware that the anonymity of the Web may lead to some bizarre responses because it's so easy to type something and send it without giving serious thought to your message. People occasionally use anonymity as an opportunity to sound off on issues that may not relate directly to the topic at hand. For example, a computer programmer told me he wrote a short questionnaire to get feedback for online help. Someone took the opportunity to rant and rave that the company's stock had gone down dramatically.

Following are some tips for sending a questionnaire through the mail:

- Send it with a letter that explains who you are, the purpose of the questionnaire, and how you'll use it.
- Always give a specific date by which you need the questionnaire returned. *"As soon as possible (ASAP)" isn't a date.*
- Include a stamped, self-addressed envelope to make it easy for respondents to fill out the questionnaire and return it.

When you prepare a questionnaire for respondents to answer online, leave enough room in your fields for appropriate answers. And provide a send button or e-mail address to make it easy for people to respond. The easier you make it to respond, the more responses you get.

Designing the Form

To get the most bang for your buck, keep the form simple and brief. You want respondents to give you lots of information with minimum effort. Here are some nifty tips on preparing the questionnaire:

- **Start with an opening that lets the respondents know this questionnaire is for their benefit, not yours.**

  Able Enterprises is constantly striving to provide you with the highest quality. We value your opinion and hope you'll take a few minutes to let us know what you think of this manual.

- **Word the questions so they're straightforward, not vague.**

  Straightforward: *Have you found any content errors? If you have, please identify the error and page number.*

  Vague: *Have you found any errors?*

- **Arrange the questions so those easiest to answer come at the beginning.** Some respondents don't bother to read much beyond the first few questions, so get them while you have their attention.
- **Group together items about the same subject.** This technique makes it easy for respondents to answer the questions and easy for you to tally the responses.
- **Provide space for additional comments.** If you're doubtful whether you addressed all possible issues, give the respondents a chance to write their own comments. You'd be surprised at how much you find out.

  Please let us know how we may make this manual more valuable.

  We welcome additional comments. Please write on the back if you need more space.

- **Deal with confidentiality.** If the questionnaire may prove embarrassing or the results need to be private, clearly assure the respondents that results will be confidential.
- **Let the respondents know how to return the questionnaire.** If you're not on hand to collect the form (such as at a seminar), give a fax number or e-mail address or provide a self-addressed, stamped envelope.

When you design an electronic questionnaire, limit the form to just a few pertinent questions. A computer screen doesn't have a lot of real estate, and you don't know how much viewing area your users have. Some may not scroll beyond the first screen.

Posing the Questions

A questionnaire is only as good as the questions it asks. When you don't word a question properly, answers lead to misinterpretation and skewed results. Test the wording with people who don't have a vested interest in the results. Doing so gives you honest perspectives.

There are two types of questioning techniques: closed-ended and open-ended. Some people respond well to one type and not the other, so a balance of the two may yield the best results.

Closed-ended questions

Closed-ended questions ask respondents to select from predefined answers that are closest to their viewpoints. Questions may be yes or no, true or false, multiple choice, or a sliding scale. Closed-ended questions are easier to answer and tabulate than open-ended, but they don't allow respondents to elaborate.

Yes, no, and in between

If you ask yes-or-no questions, give respondents a chance to opt out or respond with a middle-of-the-road answer. Following are a few examples:

Would you recommend this product to others?

❑ Yes

❑ No

❑ Not sure

Was the content appropriate for you?

- ❑ Yes
- ❑ Somewhat
- ❑ No

Multiple choice

When you'd like more elaboration, consider multiple-choice questions. The following example gives respondents a number of choices:

To be responsive to the phone inquiries of our customers, how quickly do you expect answers from customer support?

- ❑ Immediately
- ❑ Within one hour
- ❑ Within two hours
- ❑ Within a day
- ❑ Other ________________

Sliding scale

A sliding scale, shown here, gives respondents a chance to reply within a range from high to low, good to bad, or whatever is appropriate.

	Excellent	Good	Average	Poor
Organization				
Appearance				
Ease of use				
Completeness				
Quality of examples				
Meets your needs				
Overall				

If you use a sliding scale that's numbered, let respondents know which end of the scale is high and which is low. You might include the following:

1 = completely satisfied

5 = not at all satisfied

Avoid overlapping ranges that may cause confusion. In the following "don't use" example, you wouldn't know which category to select for the numbers 10 or 20.

Use: 1–10, 11–20, 21–30

Don't use: 1–10, 10–20, 20–30

Open-ended questions

Open-ended questions let respondents answer in their own words. This approach is useful because you can get a lot of feedback from explanations. Some respondents, however, don't take the time to answer open-ended questions. That's why a mix of closed- and open-ended questions strikes a balance.

The following question asks respondents for an explanation. This provides more information than a simple *yes* or *no*. (If the answer is *no,* you certainly want to understand why in order to eliminate the problem. If the answer is *yes,* just bask in the glow.)

Would you recommend this seminar to others? Please tell us why you feel this way.

Learning from the Results

When the results of the questionnaire aren't glowing or don't give you the answers you expect (or at least hope for), learn from the responses. Always regard disapproving comments as a way to reach a higher level of excellence.

I offer *Energize Your Business Writing* workshops and constantly tweak the content of the workshop based on constructive comments from participants. My participants teach me a lot and are in part responsible for the high quality of my workshops. Example 11-1 is the evaluation form I use with close-ended and open-ended questions.

Evaluation for *Energize Your Business Writing*

Name and title: ______________________ Telephone: __________

Company: ______________________ Business Unit: __________

Company Address: ______________________

I'm always looking to improve my workshops and welcome your comments. Thank you.

Workshop Content *I got the tools to:*	**Excellent**	**Good**	**Fair**	**Poor**
1. Write documents that get attention				
2. Help my reader find the key issues quickly				
3. Express my ideas more clearly				
4. Karate chop through writer's block				

Instructor	**Excellent**	**Good**	**Fair**	**Poor**
1. Knowledge of subject				
2. Confidence				
3. Enthusiasm				
4. Overall performance				

Productivity results:

1. As a result of this workshop, I expect to cut my writing time. **Yes** **No**
2. If yes, by how much? **50%** **40%** **30%** **20%** **10%**

Comments: What I found most valuable about the workshop. Any other comments?

____

____

____

Most of my business comes from referrals; therefore, I'd appreciate your letting me know of anyone who may benefit from this workshop.

Name and title: ______________________ Telephone: __________

Company: ______________________

Address: ______________________

Sheryl Lindsell-Roberts
(508) 229-8209

Example 11-1: Evaluation form for my business writing workshop.

Chapter 12

I Came, I Spoke, I Conquered

In This Chapter

- Evaluating your audience
- Knowing what to say and how to say it
- Creating visual aids and handouts that are appropriate for your audience
- Using a checklist to prepare for your presentation

> *Video [television] won't be able to hold onto any market it captures after the first six months. People will soon get tired of staring at a plywood box every night.*
>
> —Darryl F. Zanuck (head of 20th Century Fox Studios), 1946

Many people would rather undergo root canal surgery than get in front of a group and deliver a paper or give a presentation. Look at the bright side. When you give a presentation, you have an opportunity to shine and grow professionally. As a result, you can advance in your career within your company or with other companies. You may make a presentation for any of the following reasons (or more):

- Represent your research and development (R&D) group by demonstrating a new process
- Turn a research paper or report into a presentation
- Appeal to management for an increased budget in order to further your research
- Run a training session

Delivering a paper

Delivering a paper doesn't refer to the paper carrier who totes *The Wall Street Journal* to your doorstep each morning. It refers to studies, articles, or reports presented at professional conferences, conventions, or meetings. These presentations are made by scientists, engineers, physicians, or other technical professionals who are often asked to speak on a topic in which they're experts. These findings are first compiled in written form and then reported at a professional conference or convention.

Make sure that your technical paper is well thought out and carefully planned. A technical paper must be presented in logical order and accomplish the following:

1. Define the problem
2. Set the stage
3. Explain a process
4. Share the findings
5. Consider the broad implications

You present a paper orally, relying on notes, slides, or any other visual aids that can increase your audience's understanding of the topic.

Getting to Know Your Audience

Have you ever wondered why some entertainers end their shtick with, "Thank you, you've been a wonderful audience," while others seem to ponder what went wrong? The answer is quite simple. Wonderful audiences aren't born; they aren't an act of God; and they don't happen by accident.

Wonderful audiences are the result of the hard work of scriptwriters, comedy writers, and songwriters who take the time to understand the audience. You too can write a dynamic presentation and have a positive effect on your audience. But first you must take the time to get to know who makes up that audience.

Get to know your audience intimately by using the Technical Brief in Chapter 2 and on the Cheat Sheet in the front of this book. Then go one step further and answer these questions about the members of your audience:

- **What do they know about you?** Will they view you as credible or must you establish your credibility? If you must establish it, you have about two minutes in which to do it. You may start your talk with an accomplishment. Or place your biography in an obvious place so the audience can look at it before you start.
- **What do they know about your topic?** Do they have any preconceived ideas that make them friends or foes? For example, if you're disputing a popular theory or medical finding, they may be adversarial. You can disarm them by saying, "I know that many of you may not immediately agree, but . . ."

- **What motivated them to attend your presentation?** They may be motivated because you're the speaker, they have great interest in the subject matter, or their managers twisted their arms.
- **What are their objectives?** They may be there to gather information or to make personal contacts with their peers.

Getting Ready for Prime Time

When preparing for your presentation, determine five key points and subpoints that zero in on your topic. Expand each key point and subpoint as fully as you can by using the brainstorming and outlining tips discussed in Chapter 3. After you expand the points to their fullest, sequence them in the order in which they make most sense for your audience. You can find a variety of sequencing methods in Chapter 5.

Practice makes perfect

Practice! Practice! Practice! Even though you use notes or a fully written speech, you must sound *conversational* and make eye contact with your audience. Following are some practice tips. I don't guarantee that they'll make you a spellbinding speaker, but you certainly will make a good impression and deliver your message in the best possible way.

- Practice in front of a mirror or in front of peers who will be constructive.
- If possible, practice in the room in which you'll make your presentation.
- Practice with your visual aids, not just your notes.
- Tape-record your talk to hear how you sound. (If you have the capability, ask someone to videotape you to see how you appear to your audience.)

Timing is everything

Be aware of the time allotted for your talk. Remember that an eight-page, double-spaced manuscript takes about 15 minutes to present — without visuals. When you speak before your audience, you generally speak more quickly than you do in front of a mirror.

Get comfortable with your environment

If you're able to, check out the room beforehand. Make sure that you have all the equipment you need, including a podium, audiovisual equipment, and name tents. Find out whether you're presenting near an airport, a main highway that has trucks whizzing by, or an area where heavy construction equipment is being used. You may not be able to do anything about distracting background noises, but you must prepare for them. For example, build in a little extra time if you have to pause because of loud noises.

Conveying Your Message with Confidence and Competence

Use language that's clear, concise, and conversational. Keep it short and simple. Use positive words and the active voice. Be sensitive to word associations, sarcasm, and sexist language. For more information about using the right tone, check out Chapter 6. Here are three specific ways to get the right tone and stress your message:

1. **Phrase your sentences so that they're strong and have impact.** (Check out Chapter 6 to find out how you can use the active voice to accomplish this.)

   Strong: Financial planners believe that the market will continue to rise.

   Weak: There's a belief among financial planners that the market will continue to rise.

2. **Use your voice to highlight the words or phrases you want to stress.** Use boldface or a marker on your notes to indicate where you want to add stress with your voice. Notice how that's done in the following sentences:

   The goldfish is in the sink. (Simple statement of fact. Nothing is stressed.)

   The ***goldfish*** is in the sink. (As opposed to the shark.)

   The goldfish ***is*** in the sink. (In case you doubted it the first time.)

   The goldfish is in the ***sink.*** (As opposed to in the bathtub.)

3. **Use statements and pauses.** For example, make a bold statement. Pause. Say, "Think about that for a moment." Pause again.

Use repetition strategically

You don't want to be repetitive to the point of boring your audience, but you may use repetition strategically to give strength to key ideas. Notice how the example that follows creates a lasting image in the audience's mind.

> **Strong:** Why should we adopt this policy? We should adopt it because it will give us the competitive edge. And we should adopt it because it will give us a 25% profit.
>
> **Weak:** Why should we adopt this policy? Because it will give us the competitive edge and a 25% profit.

Leave these phrases at the door

Table 12-1 shows phrases to omit from your talk. Your audience may perceive them incorrectly (or correctly).

Table 12-1 Avoid These "Speaker Says" Phrases

Speaker Says . . .	*Audience Perceives . . .*
I'm really not prepared.	Why should I waste my time listening to you?
I don't know why I was asked to speak here today.	Am I being victimized by someone's poor judgment?
As unaccustomed as I am . . .	Thanks for sharing that. I should've stayed away.
I won't take up too much of your time.	The speaker protesteth too much. This is going to put me to sleep.
I don't want to offend anyone, but . . .	Oh, no. Here comes an insult.
Have you heard the one about . . .	Jerry Seinfeld he's not.
Just give me a few more minutes.	It's already been too long.

Don't start your talk with "Good morning. For those of you who don't know me, my name is [name]." Whenever I hear that, I always wonder, 'What's your name for those who *do* know you?' Simply say, "Good morning. I'm [name]."

On the foreign front

It's a small world and shrinking quickly. International travel is commonplace. If you have occasion to speak before a foreign audience (even if they're foreigners visiting the United States), you must display international savvy.

Here are a few suggestions for speaking to people from foreign countries or cultures:

- Start your talk by expressing your sincere honor at being able to address the group.
- Deliver a powerful line or phrase in the audience's native language. If you don't know the native language, ask someone you trust to translate your phrase.
- Be aware of any current events that surround the country or culture and be sensitive to those issues.
- Cite a well-known person from your audience's country or culture. (Make sure that person is someone your audience admires.)
- If you're talking about measurements, use metric terms. The United States is one of the few places in the universe that doesn't use the metric system. You see a listing in the appendix.
- Never (inadvertently) insult your audience with cute remarks that are social blunders. For example, if you're speaking to people from Greece, don't announce that "in America, all the diners are owned by Greeks." That's not a compliment to them; it's an insult.

Choosing Appropriate Visual Aids

This section focuses on transparencies and slides because they're the most commonly used visual aids. Table 12-2 lists the advantages and disadvantages of these and other options. The choice of visual aids may be dictated by your company, the industry, your audience, your budget, the complexity of the data, or the available resources. When the choice is yours, base your decision on these issues:

- **Common sense.** For example, if you're trying to convince your manager that you need additional funding to continue your R&D project, it's probably best to submit a report to the manager. However, if you're trying to convince a high-level committee that you need the funding, you may want to prepare a presentation.
- **Reusability.** If the presentation is to be used over and over again (by you or anyone else), you may be able to justify the cost of a slide or video presentation.

Table 12-2 Pros and Woes of Visual Aids

Medium	*Pros*	*Woes*
Marker or white board	Easy to find and use. Inexpensive. Informal.	Low visual impact. Boring. Limited audience size. Hard to face audience and write.
Flip chart	Good for fewer than 20 people. Easy to find and use. Inexpensive. Informal. Good for audience interaction. Can be pasted up around room for reference.	Low visual impact. Flipping back and forth can be distracting. Hard to face audience and write.
Transparency (also called "overhead")	Good for 75–100 people. Quick to prepare. Inexpensive. Flexible for tailoring presentation. Good interaction with audience.	Switching transparencies may be distracting. Projectors can block audience's view. Photographs don't copy well.
Slides	Good for several hundred people. Higher quality than transparencies. Formal. Projectors are easy to carry. High visual impact. Long shelf life. Good for copying photographs.	Expensive. Darkened room prohibits interaction with audience. Can't be redone easily.
Video	Unlimited size audience. Very high visual impact. Very formal. High interest level.	Audiences focus on video, not speaker. Costliest medium. Takes time to produce.

Preparing transparencies and slides

Most of today's software lets you design and create transparencies, 35mm slides, and handouts. Many software programs have built-in drawing tools, automatic design features, and clip art. Be as creative as you dare and pay attention to the power of visual impact.

In Example 12-1, you see a plain vanilla transparency that's ho-hum at best. Example 12-2 transforms it. Although you see it here in black and white, the actual transparency has white text on a navy blue background and uses one large red check mark to replace the bullets.

How Orders Are Allocated

After sorting orders by start date and priority, the allocation program determines which orders to allocate by considering the following:

* Start date
* Order status
* Order type
* Issue method
* Item status

Example 12-1: Delivers the message but is glitzless.

How Orders Are Allocated

After sorting orders by start date and priority, the allocation program determines which orders to allocate by considering the following:

Start date
Order status
Order type
Issue method
Item status

Example 12-2: Delivers the same message with glitz.

Consider putting frames around your transparencies or separating them with paper. Otherwise, static electricity causes them to stick together. The result can be worse than wool socks coming out of the clothes dryer.

Awesome organization

Check out Chapter 3 for a variety of ways to organize your presentation. This is based on the impact you want to have on your audience. Here are some hints:

- Start with a summary or a brief overview. State your key issues up front.
- Clearly state the problem/need.
- Solve the problem/need by making recommendations.
- Use transitions between your topics.
- Back up your recommendations. Second guess your audience and answer questions or objections they may have.
- Tactfully push for action. (Don't underestimate your power of persuasion.)
- Summarize your main points. Repeat the conclusions you've drawn.

When you have a favorable or neutral audience, place the key issue first. When you have an opposing audience, build up to the key issue.

Guidelines for text

Designing great visuals is an art you can learn to do simply and creatively. One key is to use your visuals to display the *highlights* of your talk — not every word. Otherwise, you come across as a boring talking head. Remember that people attend your presentation to hear you, not to watch you read.

The following guidelines can help you prepare visuals that hit the mark:

- **Convey one point per visual.** The point may be what, where, how much, or any single issue you want to communicate.
- **Use fonts appropriately.** Use a 24-point font for the headline and 18-point for the text. (Your visuals must be easy to read, even from the worst seat.) Use uppercase and lowercase, even for the headlines.
- **Select colors carefully.** You don't want your audience to squint at white text on a pale green background.
- **Limit your text.** Limit each visual to between five and seven double-spaced lines of text. Use bulleted or numbered lists where appropriate.

Develop a template to establish a consistent style. Doing so unifies your message. And always strive for visibility, clarity, and simplicity.

Guidelines for graphics

Studies continue to show that people assimilate graphics more quickly than text. So, if a table or chart gets your point across more quickly than text, use it. For more information about preparing visuals, check out Chapter 5. Following are a few tips:

- **Limit data to what's absolutely necessary.** Never put two graphs on one visual. If you have two graphs, use two visuals.
- **Label axes, data lines, and charts for easy understanding.** For example, the vertical data line may be sales in increments of thousands, and the horizontal data line may be calendar months or years.
- **Keep chart lines thinner and lighter than data lines.** Use lines to create structure, not to overpower the visual.
- **Use color to punctuate your message.** For example, if you use blue bars in a graph, you may want to strategically create a red bar to highlight a value you want to emphasize.

Giving Them Something to Remember You By

Always give the audience a handout or "leave-piece" — something to remember you by. This may be material that supports your presentation or a paper copy of your presentation. If you leave a paper copy, consider including annotations, questions, or a place for notes. Example 12-3 shows a transparency that opens a training unit with a list of questions for the audience to answer.

Giving handouts before your presentation

Some presenters distribute handouts before the presentation and reference them throughout. This approach is appropriate for a training session or a presentation during which you want the audience to follow along. A disadvantage may be that the audience reads the handout rather than listens to you.

Giving handouts after your presentation

Some presenters save the handouts until the presentation is complete. You may want to use this approach when the handouts include data that supports your presentation. The disadvantage here is that the audience doesn't have a chance to ask questions. They may just stuff the handouts into their briefcases as they run out the door and never read them.

Never apologize for the quality of your transparencies, slides, or handouts. Redo anything you're not proud of. What you don't want to leave is a bad impression.

Checking Out Before Checking In

Before you deliver your presentation, here are a few points to double-check:

- Is my objective crystal clear?
- Did I learn everything I can about my audience?
- Are my visuals informative and pleasing to the eye?
- Have I practiced my presentation in front of a mirror or before my peers?
- Have I anticipated some difficult questions? Am I prepared to answer them?
- Have I confirmed the date, time, and place of the presentation a week in advance?

Check out the facilities as soon as you arrive. A friend of mine did a training session in the conference room of a hotel. He arrived early and found that the room didn't have the phone jacks he needed to hook up his computer. He wound up stringing a phone line across the lobby while hanging from the grids in the suspended ceiling and almost fell to the floor. Worse yet, he may never be able to show his face in that facility again.

Inventory Training Guide *1*

Unit 2: Inventory Transactions

Goals of Unit 2:
At the end of this unit, you should be able to explain the

- ❑ Difference between adjustments and move
- ❑ Miscellaneous Receipts and Issues window
- ❑ Difference between location to location moves and transfer orders
- ❑ Receiving and issuing processes

Here are some questions you might think about as we go through this section. Answer them at the end of this unit.

1. At the very basic level, what is inventory?

2. What's the difference between an adjustment and a move?

3. When inventory doesn't arrive through normal channels, what window would you use to process the "miscellaneous" inventory?

4. What's the difference between a move code and a reason code?

(continued)

Example 12-3: Handout with transparency and applicable questions.

Chapter 13

Executive Sum-Upmanship

In This Chapter

- Realizing the importance of an executive summary
- Understanding the anatomy of a report
- Knowing what to include in an executive summary

Statistics show that teen pregnancy drops off significantly after age 25.

—Mary Ann Tebedo (member of the Colorado State Senate), quoted in *The Denver Post* (5/14/95)

The *executive summary* comes by its name very logically. It's a summary intended for executives who need a condensed version of the key elements of a lengthy, formal report in order to make timely and appropriate decisions or recommendations.

An Executive Summary Is Critical

In *Effective Communications for Engineers,* author Roy B. Hughson cited the study done by Westinghouse Electric Corporation entitled "How Managers Read Reports." The study confirms that managers read the executive summary even though they may read little else. Following is a breakdown of what managers read in a report:

- Executive Summary: 100 percent
- Introduction: 65 percent
- Body: 22 percent
- Conclusions: 55 percent
- Appendix: 15 percent

Anatomy of a report

A report is an impartial, objective, and planned representation of facts. When you write an informal report in the form of a brief memo or e-mail message, you generally include an introduction, body, conclusion, and recommendation.

When you write a formal report, you include a lot more. The following list shows the anatomy of a formal report, even though you may not use each component in every report. For example, your report may not have a list of figures or a glossary. This sample list simply shows the whole enchilada.

Front matter

- Title page
- Abstract
- Table of contents
- List of figures
- List of tables
- Preface (or foreword)
- List of abbreviations and symbols

Body

- *Executive summary*
- Introduction
- Text (including headings)
- Conclusions
- Recommendations

References

Back matter

- Bibliography
- Appendixes
- Glossary
- Index

For an in-depth look at writing rousing reports, check out my book *Business Writing For Dummies* (Hungry Minds, Inc.).

Most executives often don't have the time to read the details. Yet they make decisions or recommendations about personnel, funding, policies, or other key issues based on the information they digest in one or two pages of an executive summary. Make those pages action-packed and chock-full of critical information. Example 13-2 at the end of this chapter shows an executive summary with all the bells and whistles.

Summarizing for the Executive

You write an executive summary after you write the entire report — so don't even try to write it before you finish the report. (Think about it. You can't write a book report before you read the book.) Because the executive summary is often the only part of a report an executive reads, it's critical that you include (at the very least) the following key parts. Also include graphs, tables, or charts if they illustrate key points at a glance, as shown in Example 13-1.

- Purpose
- Findings
- Recommendations
- Background (if necessary)

Sequence your information to have the most impact on your reader

Check out No. 6 on the Technical Brief to know the reaction of your (executive) reader. Then check out Chapter 5 to find out how to sequence your document. When you anticipate that your reader will view your summary as favorable, put the findings or recommendations first. Why hide them? Here is the order in which you may present information to a favorable or neutral reader. (You wouldn't number the headings. I did it here to show the priority.)

1. Purpose
2. Recommendation or findings
3. Analysis and supporting data
4. Background

When you anticipate that your reader will oppose your findings or recommendations, you need to build up to them by leading with the background and supporting data. Here is the order in which you may present information to an opposing reader:

1. Purpose
2. Background
3. Analysis and supporting data
4. Recommendation or findings

Never introduce into the executive summary any information you didn't include in the report. This is a summary.

A graphic is worth a thousand words

A graphic (or picture) *is* worth a thousand words, so don't hesitate to create a chart or table to condense lots of text in your executive summary. In Example 13-1, you see an informal table that condenses five pages of detailed information. (Check out Chapter 5 for more about creating tables.)

Phase	Action
Design phase	• Focus on prioritizing functions and data elements based on strategic value. • Be sure e-commerce governs the design.
Development phase	• Use a SWAT team approach. • Provide for training and on-going user involvement. • Use the timebox concept as a forcing function.
Deployment phase	• Deploy this program in three incremental phases. • Establish a hot line with technicians who are qualified to make diagnoses.

Example 13-1: Table condensing five pages of text.

Use a tone that's appropriate

Following are some issues to keep in mind regarding tone and terminology. Check out Chapter 6 for a wealth of information about what is and isn't appropriate. Here are a few highlights:

- **Use technical terms cautiously.** Don't use technical terms unless you're sure that the executive reading the report is familiar with them. Not all executives have technical backgrounds, so this is where the Technical Brief is valuable to understand your reader.
- **Show a positive attitude.** This makes me think of a story in *The Art of Possibility* (Harvard Business School Press), written by Rosamond Stone Zander and Ben Zander. It tells of two shoe factory scouts sent to Africa to prospect business. One scout sends a telegram saying, SITUATION HOPELESS [STOP] NO ONE WEARS SHOES. The other sends a telegram saying, GLORIOUS BUSINESS OPPORTUNITY [STOP] THEY HAVE NO SHOES. If these headlines were presented in an Executive Summary as "Findings," which would please the executive? It's a no-brainer.
- **Use the active voice.** The active voice is stronger and more alive than the passive voice. For example, the active voice says, "Malcolm will present his findings next Friday." When you use the active voice, you place the focus on the doer of the action. When you use the passive voice, you place your focus on the action, nor the doer. The passive voice is dull and weak, as you see in the following sentence: "The findings will be presented by Malcolm next Friday."

- **Think seriously about being funny.** Incorporating humor into an executive summary (or into any part of your report for that matter) isn't appropriate. Your company isn't paying you to be a comedian.

There's a big difference between an executive summary and an abstract. An abstract is a snapshot of a long report or article. It helps the reader decide whether she wants to read the long text. Check out Chapter 9 for more information on writing an abstract.

Seeing Is Believing

Example 13-2 is the executive summary from a report I wrote for the U.S. Coast Guard. I was asked to edit the original draft of 206 pages because it was "slightly" too long. ("Slightly" was an understatement.) The document teemed with repetitive information about the Coast Guard ports and more. I streamlined the report with charts, tables, and graphs into a final document of 28 pages. I further pared the essence of the report into the following one-page executive summary by incorporating a table and a numbered list.

Executive Summary

This study documents the costs and benefits of potential U.S. Coast Guard Vessel Traffic Services (VTS) in selecting U.S. deep draft ports on the Atlantic, Gulf, and Pacific coasts.

Findings

The study indicates that the 23 study zones can be divided into three groups in terms of their relative life cycle net benefits. The following groupings are divided into areas of sensitivity:

Net benefit	Port
Positive	New Orleans, Port Arthur, Houston/Galveston, Mobile, Los Angeles/ Long Beach, Corpus Christi
Sensitive	New York, Tampa, Portland (Oregon), Philadelphia/Delaware Bay, Chesapeake North/Baltimore, Providence, Long Island Sound, Puget Sound
Negative	Jacksonville, Wilmington, Santa Barbara, Portsmouth, Portland (Maine), San Francisco, Anchorage/Cook Inlet, Chesapeake South/Hampton Roads

Approach

The following summarizes the seven steps used to gather the data:

1. Defining study zones and subzones.
2. Analyzing historical vessel casualties.
3. Forecasting avoidable future vessel casualties in each study zone.
4. Estimating the avoidable consequences in each study zone, the associated physical losses, and the dollar values of these avoidable losses.
5. Estimating the costs of a state-of-the-art Candidate VTS Design for each study zone.
6. Comparing the benefits and costs among the 23 zones.
7. Analyzing the sensitivity of relative net benefits among study zones to a range of uncertainty in key input variables.

Background

The concept of VTS has gained international acceptance by governments and maritime industries as a means of advancing safety in rapidly expanding ports and waterways. VTS communications are advisory in nature, providing timely and accurate information to mariners, thereby enhancing the potential for avoiding vessel casualties. This study builds on the experience of earlier efforts and provides the most comprehensive and quantitative analyses of VTS costs and benefits.

Example 13-2: Executive summary for a neutral reader.

Part IV
Computers and More

"Hey, Al! Do we hyphenate 'Doo Hickey'?"

In this part . . .

Online documents and the Internet have created an electronic tidal wave. Technical writers who limit themselves to writing paper documents simply can't survive. But those who are well-versed in both media can freely ride the wave.

To be an admirable technical writer, you must be familiar with both the paper and the electronic media because both are alive and well in the technical world. This part takes you through what you need to know to surf your way through the sea of electronic media — from computer- and Web-based training to using the Web for research.

Chapter 14

Doing_Research_Online.com

In This Chapter

- Surfing the Internet
- Getting chummy with your browser
- Doing searches via URLs and search engines

Just the other day I listened to a young fellow sing a very passionate song about how technology is killing us and all. But before he started, he bent down and plugged his electric guitar into the wall socket.

—Paul Goodman, American author and poet

The Internet has become a key ingredient in every sector across the globe. It's reshaping the way we speak, think about communications, do business, and conduct research. The purpose of this chapter is to demystify the mysteries of the Internet and show you how you can use it to conduct extensive research — research that may otherwise take you insurmountable amounts of time. The Internet gives you access to everything from finding parts for an old piece of equipment to the latest research on just about any topic you can imagine.

Surf's Up

In days of old, surfing was something to do in California or Hawaii. You'd hear, "surf's up," you'd grab your surfboard, and off you'd go into the mighty Pacific Ocean. "Surf's up" is still a familiar cry in coastal areas, but new technology has brought another meaning to the expression. In techno terms, *surfing* means "searching the Internet."

Weaving a Web

There's a spider that crawls through our offices and homes. It's so seductive that it ensnares us. Perhaps that's why it's called the Web (a shortened version of the World Wide Web). The Web is the star attraction of the Internet because it gives you access to computers and information all over the world.

The Web is growing like Topsy. It consists of billions of connections called *hyperlinks* or *hotlinks* (a pointer from text or a graphic on which you click your mouse) that let you jump from one page or site to another. All you do is choose a hyperlink, and you open a page or file with the knowledge you want.

Surfing is as easy as driving a car

If you haven't used the Internet for research because you think it's too complicated, think back to when you learned to drive a car. All those gadgets. All those mirrors. All those choices. (If you learned on a shift, you'll really relate to this.) If you pressed down on the gas pedal before you coordinated the brake and clutch, you crunched and lurched — not to mention that you ruined your transmission. However, after one or two tries, you got the hang of it, and driving became second nature.

Although the Internet isn't something you can wrap your arms around like a steering wheel, learning to cruise around is much the same thing. Like driving, surfing the Internet is a matter of just sitting down and doing it.

One big cyber-library

Now that you have access to a computer, you no longer have to visit your local library to conduct research or look up a reference. Instead, you can use the Internet, which you may think of as a colossal card catalog for a global library — complete with nearly unlimited knowledge. With the click of a mouse, you can make airline reservations; visit a UFO site; order books, CDs, or videos from an e-bookstore; take a whirl through the Louvre in Paris; buy a Pez dispenser; or get information on almost any subject you can imagine. It's all there for your surfing pleasure.

Just Browsing

This chapter doesn't cover how to get online. The way to do that varies according to the way your computer is set up and whether you're on a network. If you're a newbie, offer a coworker a box of chocolate chip cookies, and he'll help you out.

The primary purpose of a browser is to let you view HTML documents. You must have a browser to use an Internet, intranet, or extranet. The browser is computer software that lets you search for information on the World Wide Web. (It's also called the Web or www, which has a lot more syllables.) The two popular browsers as of this writing are Microsoft Internet Explorer and Netscape Communicator. Once you're in the browser, you're free to cruise around. Following are two ways to find what you're looking for: URLs and search engines.

"You are el" and you are there

An Internet address is called a URL, which stands for Uniform Resource Locator. It's pronounced by saying the letters "you are el." Here's what a URL may look like, followed by a description of each part:

```
http://www.lindsell.com
```

- **http** stands for *hypertext transfer protocol.* It alerts the browser that the URL you're looking was created by using hypertext markup language (HTML).
- **www** stands for World Wide Web. It's actually the name of a computer server, and it tells the browser that the site you're looking for is on a computer named www.
- **lindsell** is the domain name.
- **.com** is an extension that stands for a commercial site. (Check out the sidebar "Connecting the dots" later in this chapter to find out about other extensions.)

Hot sites

Following are several URLs you may find useful for technical writing. They're online magazines, technical writing organizations, and related organizations.

- **American Society for Training and Development:** `www.astd.org`
- **Good Documents:** `www.gooddocuments.com`
- **IEEE Professional Communication Society:** `www.ieeepcs.org`
- **International Society for Performance Improvement:** `www.ispi.org`
- **Project Management Institute:** `www.pmi.org`
- **Society for Technical Communication:** `www.stc-va.org`
- **Society of Documentation Professionals:** `www.sdpro.org`
- **Web Review:** `www.webreview.com`

Searching for the Holy Grail

If you don't know the URL for a site you want to search, use a search engine. You do that by entering a keyword or words and clicking a button to start the search. Then, through the miracle of modern technology, you see on your screen a listing of hyperlinks to sites that the search engine found.

Search engines rank the sites in the list based on a variety of criteria. These include how many times your keyword appears in the document, whether the keyword appears in the title, how early the keyword appears in the text, and so on. So your best hits generally appear at or near the top of the list. For more details on Web research, check out *Researching Online For Dummies,* by Reva Basch and Mary Ellen Bates (Hungry Minds, Inc.).

Not all search engines are created equal

None of the search engines cover the billion-plus searchable pages on the Web. If you don't find what you're looking for on one search engine, try one or two others. Although you may get some overlapping results, you may see variations. Following are some of the popular search engines and their URLs:

- **Ask Jeeves:** www.ask.com
- **Deja:** www.deja.com
- **Excite:** www.excite.com
- **FAST:** www.alltheweb.com
- **GO.com:** www.go.com (formerly Infoseek.com)
- **Google:** www.google.com
- **HotBot:** www.hotbot.com
- **Lycos:** www.lycos.com
- **Northern Light:** nlsearch.com or www.northernlight.com
- **Yahoo!:** www.yahoo.com

To search for information from the widest variety of appropriate sites, download a free program called Copernic 2000. This program sorts the results of the search and stores them on your hard drive. You find Copernic 2000 at www.copernic.com.

Fine-tuning your search

You can use a number of search methods. Always try to make your search as narrow as possible so you don't wind up with long, unmanageable lists. Following are basic and advanced ways to conduct a search.

Searching 101

Most search engines make searching really simple. You type a word or phrase in a box and click the Search or Go button. (Some search engines may use other words, but the button is obvious.) You may search for the phrase, click the button, and get a list of hyperlinks that contain those words as a unit.

The following examples take you through a series of screens for a search in the search engine Yahoo!. In Example 14-1, I entered the search words *lymphoid cells* and clicked the search button. In Example 14-2, you see the results of that search listed as hyperlinks to various sites. When I clicked the second hyperlink (the first was a photo), the screen displayed the page you see in Example 14-3.

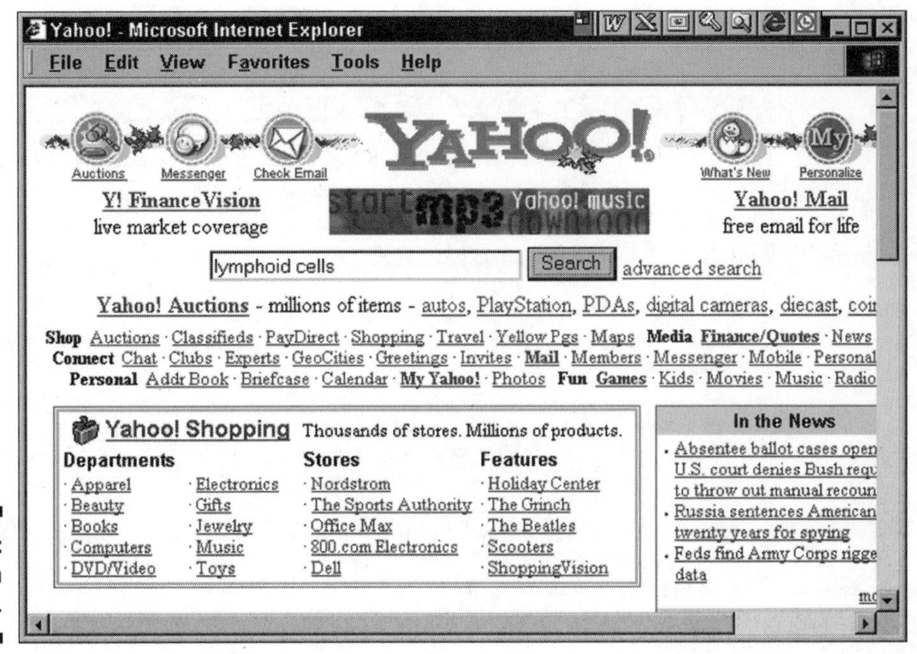

Example 14-1: Let the search begin.

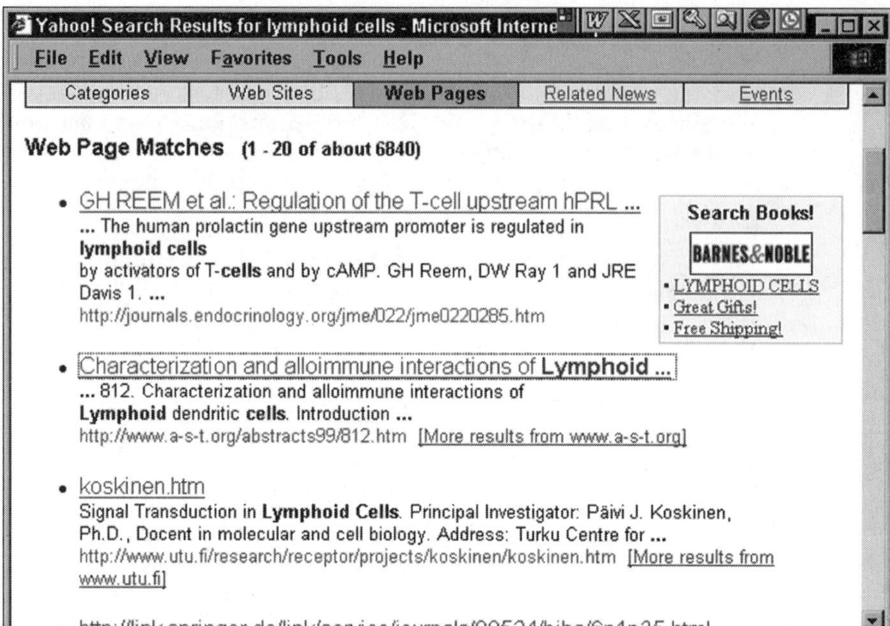

Example 14-2: List of hyperlinks.

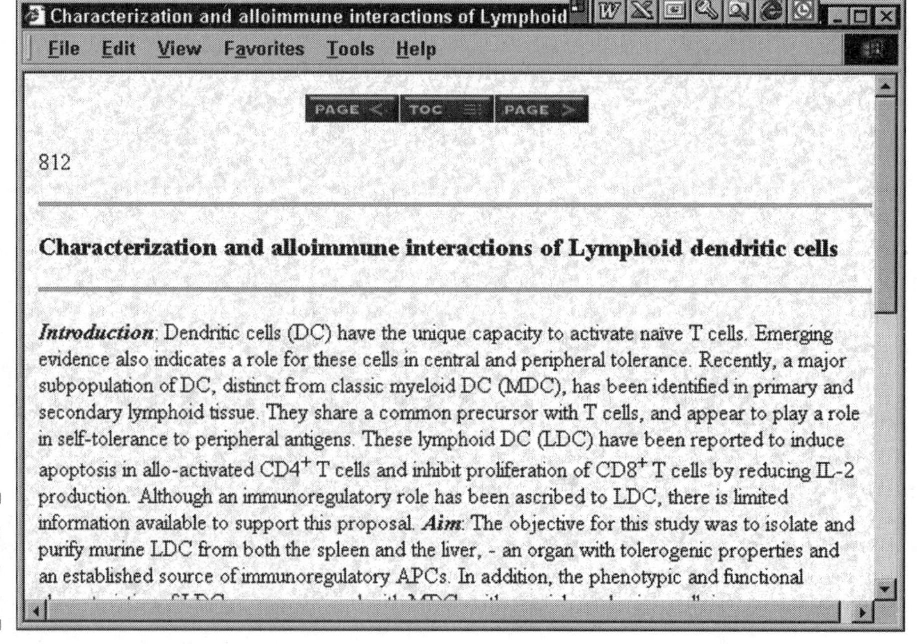

Example 14-3: The envelope, please.

Advanced Searching 102

Some search engines do a mediocre job when searching for a single word or phrase. Whether you conduct a basic or advanced search, you see results similar to the examples on the preceding pages. The difference with an advanced search is that you make your search more specific. Here are several advanced search methods you may use. Try a few and see what works for you.

Boolean search

The mathematician George Boole developed a system of logic to limit the results of your search to very specific information by using the words *OR, AND,* and *NOT* in all capital letters. Following are some of the Boolean search methods shown by example. (With some search engines, you may have to click on Advanced Search.)

- **Einstein OR Hawkins:** When you use the word OR to separate your search components, you get a list of hyperlinks where either the name *Einstein* or the name *Hawkins* appear.
- **science AND (zoology OR ecology):** When you use the word AND to separate your search components, you get a list of hyperlinks where the words *science and zoology* or *science and ecology* appear. The site won't contain *zoology* and *ecology* because you put "OR" in parentheses.
- **hardware NOT (monitors OR printers):** When you use the word NOT to separate your search components, you get a list of hyperlinks where the word *hardware* appears. The words *monitor* and *printers* don't appear anywhere in the text.

Other syntax

Here's how to narrow your search by using other syntax:

- **Quotation marks (" ") focus your search on a particular phrase.** This treats the words inside the quotation marks as a phrase:

  "Boston Aquarium"
- **The plus sign (+) indicates a specific word or words you want in your listing.** In the following example, both *mutant* and *secretion* would appear:

  +mutant +secretion
- **The minus sign (–) indicates specific words you don't want in your listing.** In the following example, mutant would appear; secretion wouldn't appear:

  +mutant–secretion

- **The asterisk (*) searches for the root of a word that may have different endings.** In the following example, you may get listings for *biology, biologist,* and *biological:*

  biolog*

- **The percent sign (%) searches for any word that has variations in spelling.** In the following example, *criticize* is the American spelling, and *criticise* is the British spelling:

  critici%e

Understanding Cyberbabble

Sometimes you get weird messages that really don't mean anything when you see them on the screen. The computer often talks in cyberbabble. Here are a few you may see and what they're trying to tell you:

- **404 not found.** In plain English, this message generally means one of two things:
  - The link is no longer available, in which case you may search for the information through your browser. If you don't find the site through the browser, the site may have vanished into cyberspace.
  - The URL isn't correct. Try typing it again.
- **The server does not have a DNS entry.** It's likely that you typed the URL incorrectly.
- **The server may be down or unreachable.** The site may be overloaded, which means too many people are trying to get to the site at the same time. Or the site may be offline for maintenance, in which case you may try again at another time.

Outlaws in This New Frontier

This new frontier of cyberspace offers a new and exciting way of doing business and conducting research. However, watch out for cyberbandits who create havoc by creating viruses that incapacitate systems and unscrupulous folks who take the liberty of "stealing" information from the Internet by cutting and pasting it into their documents.

Connecting the dots

Perhaps you're confused by all the "dot-whats" at the end of a Web address. Although .com is the most common Web extension, there are many others, and more are coming on the cyberscene. Following are extensions you commonly see and the type of sites they represent:

- .com (commercial site)
- .org (nonprofit site)
- .edu (educational site)
- .gov (government site)

Adding to the confusion are country-specific extensions, such as the ones that follow:

- .ar (Argentina)
- .at (Austria)
- .au (Australia)
- .be (Belgium)
- .br (Brazil)
- .ca (Canada)
- .ch (Switzerland)
- .cn (China)
- .de (Germany)
- .dk (Denmark)
- .es (Spain)
- .fi (Finland)
- .fr (France)
- .ie (Ireland)
- .il (Israel)
- .is (Iceland
- .it (Italy)
- .jp (Japan)
- .kr (South Korea)
- .kp (North Korea)
- .lu (Luxembourg)
- .mx (Mexico)
- .nl (Netherlands)
- .no (Norway)
- .nz (New Zealand)
- .pl (Poland)
- .pt (Portugal)
- .se (Sweden)
- .us (United States)
- .tw (Taiwan)
- .uk (United Kingdom)
- .za (South Africa)

Virus alert

When you download anything from the Internet, your computer runs the risk of catching a virus. Always make sure that you have a virus checker running on your computer to alert you to any potential problems. Two popular virus checkers available for Mac or Windows operating systems are Norton AntiVirus (www.symantec.com) and McAfee Virus Checker (www.mcafee.com).

If you use either of these virus checkers, you can disinfect files you download from the Internet. After you purchase antivirus software, you can download upgrades from the Internet (at no charge) for the constantly mutating, harmful invaders. Be sure to do that regularly because new viruses are always a threat.

Copy, paste, and violate

Most sites let you copy information so that you can paste it into your word processing software. Be cautious, however, of using any of the information verbatim. The same copyright laws that apply to paper documents apply to electronic documents.

Chapter 15

Sights and Sounds

In This Chapter

- Designing an electronic page
- Using color appropriately
- Creating graphics to download
- Using the power of sound
- Incorporating animation, video clips, and virtual reality

Who the hell wants to hear actors talk?

—H. M. Warner (one of the Warner Brothers), 1927

Don't be at the *bleeding edge* of technology. When computers acquired color capabilities and multiple font choices, computer screens looked like Times Square on New Year's Eve. Users suffered from sensory overload; some still do. Too many fonts make documents look like ransom notes. Hideous colors make them look like circus posters. Don't create an overpowering effect just because you have the capabilities to do so.

This chapter isn't about designing Web sites, because that's a category unto itself. Besides, Web sites are generally posted by Web designers, not technical writers. This chapter is about using sights and sounds to enhance your electronic documents such as tutorials, computer-based training, Web-based training, and the like.

Even if you don't have the expertise to post your document and you work with a Web designer for that purpose, you're still the one who calls the shots when it comes to text, graphics, fonts, color, audio, video, and the like.

Basics of Electronic Page Design

Think of some of the electronic documents you've seen. If they don't have visual impact, you may not look beyond the opening screen. Check out Chapter 5 to learn about ways to add visual impact to your documents. Here are some highlights as they relate to electronic documents:

- **Quiet space:** *Quiet space* (also known as *white space* or *blank space*) provides a resting place for eyes and breaks information into manageable chunks. It's the electronic equivalent of white space on a printed document.
- **Headlines:** Use descriptive headlines to call out items of importance. Notice the way newspapers use headlines to draw your eye to key information.
- **Bulleted and numbered lists:** Use bulleted and numbered lists just as you do in printed documents. They make critical information stand out.
- **Charts, tables, and figures:** A picture is worth a thousand words in electronic documents as well as in printed documents. Use them to give readers information at a glance.
- **Indentations:** Too many indentations cause unnecessary complexity and confuse users. However, you should indent to show subordinate text. If you need to, break your text into several topics.

Color My World

We're so used to being surrounded by color that we tend to take for granted the effect colors have on us at the conscious and subconscious levels. For example, if you walked into a doctor's office that was painted red, imagine what your reaction would be. Perhaps you'd imagine that blood was smeared over the wall. Check out Chapter 5 for an explanation of how colors influence us.

Using colors effectively

People don't often view black-and-white computer monitors any more than they view black-and-white television sets. Black-and-white does have its place in photography, but seeing electronic documents in color is much more exciting in most instances.

Colors on your electronic document must work well with each other as background, text, graphics, and navigation tools. Check out other people's electronic documents and ask yourself the following:

- Do the colors contribute to the site or detract from it?
- Do the colors work well together or conflict with each other?
- Do you notice that the color coding has a purpose?

Avoiding color chaos

Color doesn't substitute for a good design. Following are things to avoid:

- Colors that are too similar or too bright don't display well. For example, I viewed a tutorial that had a shocking pink background with bright yellow text. I ran to get my sunglasses and still couldn't look at the screen.
- Users won't be able to print white and light yellow text. Don't use those colors.
- Black backgrounds tend to be very harsh.
- Too many colors are distracting and make users feel as if they're looking through a kaleidoscope.

Always err on the side of simplicity. Glare or background light may saturate the display and make certain colors look washed out. And older laptops may have one level of brightness and distort a wide array of color.

Graceful Graphics

You need to consider many more variables with electronic graphics than with paper graphics. Speed and storage are two key issues. Human factor guidelines show that users lose interest if a page takes more than 10 to 15 seconds to download. Following are some issues worth considering:

File formats

Several factors affect the speed of the downloading process. For example, some users still have slow modems. Everything they download seems to take an eternity. Many have been known to grow old waiting for a download, and complicated graphics just add to their frustration. So either give them gift certificates for higher speed modems or think of their woes when you prepare graphics. Select a file format that's appropriate for your user and the graphic. Following are two of the popular formats:

- **GIF (Graphics Interchange Format):** These graphic files are compressed to minimize the time it takes to transfer a file over standard phone lines. You have a choice of 256 colors. This format is great for animations that would otherwise take forever to download.
- **JPEG (Joint Photographic Experts Group):** This format offers a high-quality resolution and an array of 16.7 million colors (in case 256 aren't enough). However, the trade-off for the high quality is a longer download time.

In cold storage

The method you use to store the graphics has an effect on how fast they load.

- When you store graphics in separate files, the computer must locate them in the order in which you open them. Doing so slows down the process. On the other hand, if you embed the graphics in the document, the file may be quite large and download slowly.
- Graphics that must be retrieved from a busy network or another computer's disk may take even longer.
- If the computer has to decode or decompress files, this also slows the process.

Space-saving ideas

In essence, here's why a simple graphics download may be more practical:

- You can save disk space by creating monochrome (black-and-white) graphics and displaying them on a screen that has color. They won't have the same the glitzy impact, but everything's a trade-off.
- When you have a graphic that you use several times, such as a logo, you can save space by storing it once and having it appear in several locations. For example, store one copy and reference that copy in several locations.
- When you use compressed schemes to condense graphics, they take longer to load but take up less storage.
- Eliminate parts of the graphic you don't need, such as unnecessary borders and textured backgrounds.
- Split large graphics into smaller graphics to load more quickly.

Hear Ye, Hear Ye!

Paper is mute; the only sound it makes is shuffling or tearing. Electronic documents, on the other hand, may speak, sing, beep, or moan and groan. You can amplify the sound of a foghorn, the waves breaking on shore, a Mozart symphony, or just about anything else. People like to hear certain types of noise. If you doubt this, talk to someone who worked at IBM when the company introduced the silent typewriter. It was a flop. People wanted quiet, not silence.

Provide a link so users can decide whether they want to hear the audio or view the video. These file formats require users to download browser add-ons.

What's that I heard?

To use audio appropriately, select a sound that users associate with something relevant to the topic: a car revving its engine; computer keys going clack, clack, clack; an icon buzzing (or screeching) across the screen; the murmur of conversation; squeaky doors opening and closing; the thump, thump, thump of a heartbeat. Following are some tips for using audio effectively:

- **Punctuate action.** Create a mood or help users anticipate what's coming. Think of the last mystery movie you saw. You knew something dire was about to happen when you heard ominous music.
- **Increase the user's involvement.** Use sound to heighten awareness and give users something to listen to besides the hum of a computer. For example, if you describe a train engine, use the sound of a subtle train whistle.
- **Maintain continuity.** Sound provides a transition while users wait for the computer to respond. It's also an effective way to connect a series of disparate data.
- **Create a mental theme.** Music gives us information we can't see. It can create the illusion of fun, sorrow, intensity, or pain. Think of movies you see that use music to heighten intense scenes.
- **Use earcons.** Think of audible signals as icons for the ear — *earcons.* Beeps, bells, buzzes, and the like invoke reactions. They get attention, create a warning, prompt for action, or alert to a certain condition. If your computer said "uh-oh," wouldn't that warn you of danger?

Don't rely on sound alone to communicate your message. Not all users have audio capabilities, so be careful of what some users may miss.

Choosing the right words

For the audio part of your electronic document, use language that's crisp and appropriate. Here are some tips:

- Use complete sentences. This isn't a telegram; you're not paying by the word.
- Stick to words with no more than three syllables. Longer words don't sound as natural when they're read. This tip doesn't apply to ten-syllable technical terms, however.
- Create punctuation with tone and inflection.
- Keep your sentences short and simple.

To hear how the written words sound, read the text into a tape recorder and listen carefully. If something doesn't sound right or you stumble over words, rework those parts. Additionally, using a professional to record the audio is always wise. Find someone who does voice-overs — perhaps a semi-pro from a nearby school. They often volunteer their time just for the experience.

Am I Your Type?

The primary function of words is to communicate a message. If you overuse typefaces or use them inappropriately, you diminish this function. When you use typefaces appropriately, they attract the user's eye and organize information in terms of their importance.

Serif and sans serif

The most fundamental distinction between fonts is whether they're serif, such as Times Roman, or sans serif (literally, "without serif"), such as Arial. The difference is that serifs have "feet" and "arms" on the end of strokes. Although a serif font is what we usually see in printed matter, experts claim that sans serif is easier to read on the screen.

There are variations even within serif and sans serif fonts. Before you post your document, check the fonts out on the screen to make sure they're pleasing to the eye.

Keep the user in mind

NEVER USE ALL CAPITAL LETTERS. IT'S THE EQUIVALENT OF SHOUTING AT THE USER. ALSO, ALL CAPS ARE MORE DIFFICULT TO READ THAN THE COMBINATION OF UPPERCASE AND LOWERCASE.

This is also true for italics. Don't overuse italics; save it to punch out a few key words or expressions. Italics is difficult to read when used ad nauseum.

Never underscore documents on the Internet. The underscore is the universal sign of an active link.

So You Wanna Make a Movie

In the electronic arena, it's simple to move away from static pages and create movement. Consider animation, video clips, 3-D, virtual reality, sound clips, and a lot more. Never use any sound or movement gratuitously or you detract, rather than attract. Following are special effects that work in certain circumstances:

- **Animation:** Animation can be lots of fun when used appropriately. For example, if you manufacture airplane parts, consider showing an animated airplane "flying" across the screen. One memorable animation is the Energizer Bunny that marches across the TV screen beating on his drum.
- **Three-dimension and virtual reality:** These are great ways to take visitors on a tour of an assembly plant or let viewers select the latest model of equipment you offer. Let viewers change directions and see the object from different angles and directions.
- **Video clips:** Include a video clip to demonstrate how to install or repair a piece of equipment.

Chapter 16

Computer-Based Training (CBT)

In This Chapter

- Understanding the role of the technical writer
- Getting a handle on the CBT process
- Finding out about interactivity
- Preparing modular chunks
- Addressing business issues
- Designing a process
- Preparing a storyboard
- Testing for quality

If the automotive industry had progressed during the last two decades at the same rate as the semiconductor industry, the Rolls-Royce would today cost only three dollars and there would be no parking problem because automobiles would be one quarter of an inch on a side!

—Thomas J. Watson, founder and first president of IBM

What is this thing called CBT? It's computer-based training, also known as computer-aided instruction, or a dozen or so other terms for learning by the glow of a computer screen. CBT can be anything from a true-false quiz to a stock market simulator. Although there's no official definition for CBT, it relates to documentation in the same way a screenplay relates to a novel. It's the same stuff, but you experience it in a different way.

For the purpose of this chapter, think of CBT as an umbrella under which you have Web-based training (WBT) and online help. If you think of CBT as the parent of training, WBT and online help are the offsprings — the children.

They're discussed in Chapters 17 and 18, respectively. This chapter discusses CBT distributed on a CD-ROM. Following are a few appropriate applications for CBT:

- You want to sell the CBT in a bookstore or electronically.
- Your users need the convenience of simply inserting a self-starting CD-ROM.
- Downloading graphics would take your users too long.

The Role of the Writer

The role of the technical writer in the CBT process depends on the writer's experience and knowledge of the subject. Following are several progressive scenarios:

- The subject-matter expert (SME) may write the draft and then turn the project over to the writer for storyboarding. (Storyboarding is explained later in this chapter.)
- If the writer has knowledge of the subject matter, he may write the draft and then create the storyboard.
- If the writer has knowledge of CBT software, he may work on the project from start to finish.

"C" Is for Computer-Based

Some folks want training that's integrated into another computer application. This isn't just computer-based; it's computer-rooted, much like the roots of a willow tree in a septic tank. This takes a lot more work to write, needs a lot more technical intervention, and probably takes a good deal of programming.

"T" Is for Training

Training means helping users to master a skill — a skill they perform regularly, not just a task they perform occasionally. It's a proven fact that we retain information on the basis of how we experience it. We retain

10 percent of what we read

20 percent of what we hear

30 percent of what we see

50 percent of what we hear and see

60 percent of what we say

90 percent of what we say and do

Because of CBTs interactive capabilities, *do* is a key component. Most CBT is interactive, unlike a TV screen that you just sit and watch. CBT can include online books (where the interaction is mainly menu access), quizzes, simulations, games, multimedia, and more. Interactivity is expensive to prepare, but it's a great way to learn.

Types of Interactivity

Focus interactivity on improving the users' learning experience. Preparing meaningless bells and whistles not only wastes your time but annoys users. Understand what the users need to know and how they can best learn it. Be sure to fill out the Technical Brief you find in Chapter 2 and on the Cheat Sheet in the front of this book.

Navigational aids

In some CBT, the interactivity is limited to navigational elements such as search capabilities and menus. (That may not count as real interactivity with the tough guys in technoland — you know, the ones with bad haircuts and wild eyes. They'll be quick to spurn this genre.) If you use a little creativity, however, your users may not miss the hundreds of hours you *didn't* spend writing games, simulations, and automatic diagnostics.

Know your objective

When your training objective is to get your users up and running in the shortest possible time, you may not want to enforce a beginning-to-end training experience. Every minute users are being trained is a minute they're not doing the ultimate task. You want to teach them enough to get the job done in a way that matches their learning preferences.

The best motivator for typical adult users is to do their jobs well. *Their peak point of interest is that moment when they realize they don't know something.* If your training material can answer their questions and get them back to work quickly, not only will they enjoy the training, but they're highly likely to remember what they learned the next time.

Hand holding

If your users need all the information in your training, lead them by the hand to ensure that they get from Knowledge Level A to Knowledge Level B. You do this by controlling their path through your material. One way is by linking a table of contents so that users can always pick up where they left off.

Just in time

If you're building a CBT system that aims to get little nuggets of information to users as they need them, then focus on letting them access the right nugget in the shortest possible time. Key elements are a good table of contents, an index, and search components. And if little nuggets are what you're after for a large portion of the training experience, perhaps online help will serve that purpose, rather than CBT. Check out Chapter 18 for details on online help.

Self-evaluation

Another way of promoting interactivity is to let users evaluate themselves. In Example 16-1, you see a screen on which users can respond to short-answer questions. They fill in their answers and then compare them with the correct answers. It's fast, easy, and effective and requires no special software. Users like it because there are no records and no embarrassment. Developers like it because it's easy to write.

Thoroughly interactive CBT means that your program will respond in a smart way to something the user does. How smart depends on how ambitious you are. This can be everything from tallying up correct responses in a multiple-choice quiz that may report, "You're a genius!" to letting users specify a what-if scenario with CBT responding in a unique, intelligent way.

Simulations and games

Simulations and games let your users explore and experiment so they learn by doing. These may include anything from flight simulators to performance reviews that have multiple variables and a gazillion possible outcomes. No one will quibble about the value of this kind of learning, but it's expensive to create. Unless an application package has been built especially for the development of simulations in your content area, this is way beyond the realm of CBT.

Complete las frases

Llene los espacios con la(s) palabra(s) correcta(s) para hacer completas las frases sobre la gente de Cuba.

1. La gente cubana viene del hispano [ ] o [ ]. RESPUESTAS
2. Habían unos grupos [ ]. RESPUESTA
3. Hay problemas de la [ ] en Cuba pero las condiciones son mejores en Cuba que en los otros países latinos. RESPUESTA
4. En dos pueblos de Cuba la gente es muy alegre. Un pueblo se llama [ ] y el otro se llama [ ]. RESPUESTAS
5. Algunos deportes populares son [ ], [ ], [ ], y [ ]. RESPUESTAS
6. A los jóvenes les gusta mirar las [ ] extranjeras. RESPUESTA
7. Un hombre muy importante en la historia cubana es [ ]. RESPUESTA
8. [ ] es un lugar donde se puede comer el helado. RESPUESTA

Example 16-1: Questions with a foreign flavor.

Real-life assignments

One easy way to provide a significant learning experience that's more meaningful than true and false questions is to give users actual assignments. For example, if you determine that typical users have Excel on their computers, you can write, "Click here to determine which element has the greatest influence on the bottom line." Users click, and Excel displays a spreadsheet. You see a sample of this application in Example 16-2. After some experimentation, users return to the training. You can discuss the options, and users can compare their observations with your feedback.

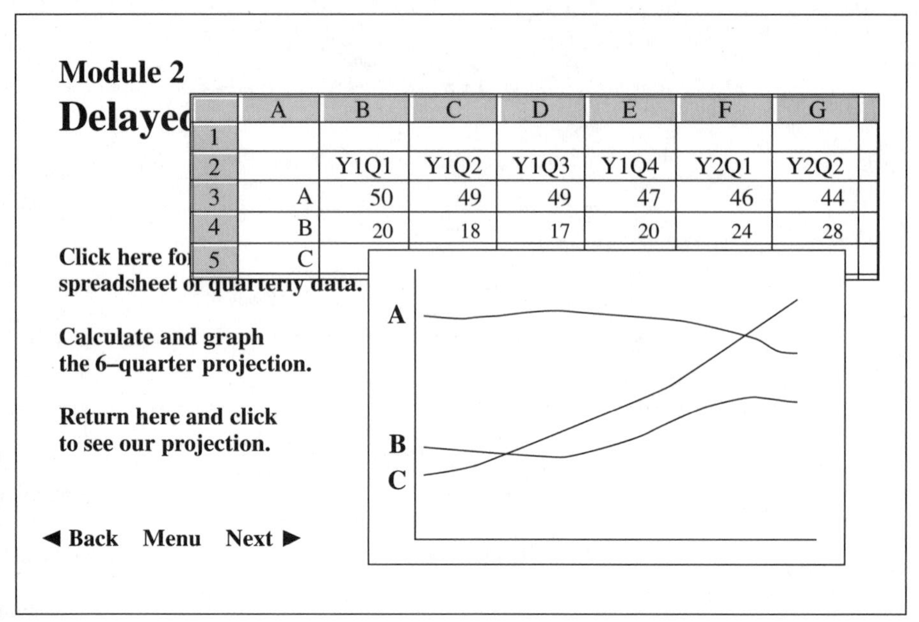

Example 16-2: Plugging in external programs.

Presenting the Learning Experience

In 1981 Dr. Roger Sperry won the Nobel prize in medicine for theorizing that the brain is divided into two parts, each performing a different function. The left side of the brain deals with logic, language, reasoning, science, and math. The right side of the brain deals with visualization and creativity. Therefore, we all learn differently, depending on how we're wired. When you prepare a learning experience, try to offer something for everyone.

Hand in hand with the decision about level of interactivity is the decision about modularity (chunks). Modularity means keeping the contents of each page focused on one thought or activity. Match the size of your information chunks with the users' training needs.

Linear learning experiences

If you're presenting a linear learning experience (taking people from Knowledge Level A to Knowledge Level B, and they have to learn everything between the opening screen and the closing credits), then you can relax into continuity. Base your chunking decisions mainly on the size of your users' thought units.

Random learning experiences

If your users will approach your fount of training with a tiny dipper and take many dips in random order, you don't have the luxury of knowing that they just learned. For randomly accessed chunks, attempt to make each chunk autonomous and completely comprehensible to the average beginner in your target audience.

Solving Business Problems

Remember that all training is designed to solve a business problem. Here's an example: My colleague Jennifer was assigned the task of training a janitorial staff on organic chemistry. Jennifer's client recycled a lot of his materials. His business problem was that the janitorial staff was having trouble keeping the *dirty* chemicals (those to be discarded) separate from the *clean* drains (where only recyclable chemicals should go). The staff members spoke a variety of languages, and Jennifer was overwhelmed at the prospect of preparing CBT of that magnitude in six languages.

When Jennifer got down to brass tacks by using the Technical Brief, she learned that the client had only one learning objective. He wanted his staff to respond to chemicals with the following behavior: "If I don't know what it is, I won't pour it down a drain." Once that was simplified, translating this simple training into six languages was a breeze.

One of the first things you and your client must agree on is why to prepare CBT.

- **What business problem is your client trying to solve?** Perhaps support technicians in the company send too many questions up to the engineers.
- **What change do they wish to make in the users?** You may find out that the support technicians should be able to answer 90 percent of the questions on the new products without calling upon the engineers.

As you delve into the second question in the preceding list, find out what users need to know and how thoroughly they need to know it. This is where the Technical Brief is helpful because you'd learn that support technicians need an in-depth knowledge of the technical information.

Big-picture learning objectives

Once you work with your client to pin down the business problem the training is expected to solve, then create measurable and observable learning objectives. Objectives may be a percentage of correct test answers or a level of competency for a task. A good learning objective may read something like this:

> *Users will be able to identify all the elements of a peanut butter and jelly sandwich and be able to assemble a sandwich successfully nine times out of ten with the assistance of a series of steps.*

To know exactly how much expertise users need, ask these questions:

- Do they need a skill level adequate for following directions to perform a simple or complex task?
- Do they need the background and understanding to handle unique situations?

Adding the virtual personal touch

Never downplay the role of personal contact via e-mail, bulletin boards, or chat rooms. Years ago I gave a pitch to a teacher about how wonderful it would be if she could capture her excellent classroom presentations in CBT modules so that students outside her classroom could benefit. She looked at me in amazement and said, "It's hard enough to reach the kids that I work with every day when I can see their faces and answer their questions. How in the world can I teach kids I've never seen, in places I've never been?"

With CBT, distance isn't a deterrent. E-mail, bulletin boards, and chat rooms provide the opportunity for users to ask questions, get clarity, and discuss options. As the author of the CBT, you probably won't be in a position to provide these things to the end user. However, you can suggest these options to your clients and include provisions for the personal touch in the training plan.

Personal contact helps users through the rough spots when they don't quite understand, don't see the need, or just get busy with other pursuits. A mentor (even a knowledgeable voice) has a positive impact on a discussion among strangers. The key objectives are to overcome confusion quickly, avoid frustration, and establish a connection with other human beings.

One big problem with personalized CBT training, however, is attrition. Without human contact and accountability, people drop out — both users and mentors.

In the 1980s, the University of Athabasca initiated an innovative program that put every user within a local phone call of a mentor (which was no small feat considering that many of the students lived in the tundra). This initiative helped to make a personal connection, and with CBT you can do a whole lot more.

Getting more complex

There's a flavor of CBT called Electronic Performance Support System (EPSS). In simple terms, EPSS proponents feel there's no sense in squeezing something into users' brains if they can get the information off the hard drive fast enough to simulate knowing it.

For example, when you call your insurance company from Cozumel to find out whether they'll cover the extraction of a fishhook from your foot, the folks at the toll-free number haven't memorized that level of detail. Their database feeds them the information so they can answer your question without losing a beat. This same analogy applies to workers assembling complex systems on an assembly line. They don't commit the procedures to memory; they watch the monitor for instructions detailing each step in the process.

Platform Independence

Version problems are a fact of life. You must know just how antique a user's computer may be in terms of the following:

- Operating system (which version of Windows, Mac O/S, or other)
- Hardware (what generation processor, how much memory, and with what kinds of drives for removable disks)
- Speed of Internet connection (if WBT)
- Version of user's software and browser

Squeezing an orange through a phone line

Of course, you can't squeeze an orange through a phone line; the orange is too fat. And you can't squeeze a huge training program down a modem connection for the same reason. When you plan your CBT, you need to know how your users expect to receive it.

- If they want the training delivered to their home over the Internet (battling Junior for modem time), it must be lean and mean, able to leap through long wires in a single minute (or in a few minutes). That means omitting multimedia presentations, which take up a lot of space.
- If you send it to the office where users have a fast connection, you can afford to put a little meat on the bones or package it in bigger modules. For this application, multimedia is very appropriate.

If you deliver the program on a CD-ROM, you'll probably have room to burn. But you need to ask the question, "Will users run the training off the CD-ROM or copy it to the hard drive?" The average home computer probably doesn't have the disk space to copy a bunch of big multimedia files. You may choose a hybrid installation so users can leave big multimedia files on the CD-ROM and put interactive portions on the hard drive where they'll run faster.

High-quality video files can burn up a lot of disk space, so don't consider a CD-ROM as a license for creating huge files.

How much techno-geekiness are you up to?

For many folks, the concept of technical documentation and the notion of CBT diverge about the time they get to multimedia. Two common assumptions are that documentation doesn't include video, and training has to have video. Putting assumptions aside, we all use what works best under the circumstances. But before you decide how much multimedia is good for your users, be sure that you know how much you're up to.

Making multimedia

Multimedia is audio/video (A/V). It can come from an animation you create, an interview with the top dog, or a videotape from your client's bottom drawer. While your lips are promising to include multimedia in your CBT, you need to have your brain ruminating on these questions:

- Does my client expect a sprinkling of multimedia or a computer-delivered TV show?
- Do I need to corral the talent to script, perform, tape, produce, and edit this A/V material?
- Do I have the hardware and software to convert standard audiotapes or videotapes into digitized files?
- Do I have software (such as QuickTime Pro or Adobe Premiere) to mix, match, and edit A/V elements?
- Do I have a software package (such as Flash) to create animation files?
- Do I have the software (such as Media Cleaner Pro) to edit and compress huge A/V files into a manageable size?

If you sense that multimedia is a mega-effort, you're right. But here are very good reasons to include A/V files in training:

- Action and drama make lessons fun and memorable.
- People like to look at other people; therefore, you foster interest by creating a human connection.

- You demonstrate complex or time-dependent concepts. (Just think of how simple it would be for Zeb [the Martian from Chapter 2] to make a sandwich if he mimicked what he saw on a multimedia presentation.)
- Audio voice-overs keep users' eyes focused on an image.

Audio is much easier to handle than video because the files are smaller and users are generally less critical. Sometimes, with a little sleight of hand, you can provide multimedia benefits without going full-bore into video production. Consider pairing an audio clip with a still image or presenting a series of still images rather than motion video.

Using WYSIWYG editors

The new point-and-click editors make life easy for CBT authors. These editors provide "What You See Is What You Get" (WYSIWYG) and eliminate the need to insert arcane tags or commands for formatting.

Designing a CBT Process

In any good architecture, the vision drives the details, and the details shape the vision. Therefore, a good outcome depends on a good design. It requires a knowledge of how people learn, and it deserves a lot of time and care. I remember one conversation with a gentleman who was inquiring about a career in CBT development. His shocking comment still rings in my ears: "I know algebra; I could write CBT about algebra."

"Oy vey!" I thought to myself. The fact that he knows algebra has no bearing on his ability to transfer the information in a meaningful way to someone else. We've all had professors who know the subject matter but don't communicate it effectively.

Following is a brief summary of issues to consider. You notice there are no numbered steps in the process. As you consider the various steps and issues, you'll find yourself rethinking your initial perceptions, clarifying objectives, and reshaping your vision. The following items are those you need to visit and revisit as you go through the process:

- **Set your sights.** Envision the solution in terms of optimizing learning, the users' expectations, the client's business requirements, and your own constraints. This includes the following:
  - Time and money constraints on design, development, testing, and dissemination
  - Life expectancy of the document and maintenance requirements
  - Distribution channel(s)
  - Level of multimedia

- Level of interactivity
- Business goals and learning objectives
- The minimal computer configuration for target users
- Available authoring tools and multimedia editing tools

- **Plan the users' experience.** Eschew the old-fashioned model of the user as an empty vessel to be filled with knowledge. Learning means interacting with the material so that it's meaningful to the users. You're designing an internal experience as much as Alfred Hitchcock designed a movie. Not quite as disturbing an experience, one would hope, but more significant for the individual. Determine the following:
  - Appropriate mood to support the purpose and appeal to the users (mood can be reflected in tone, graphics, colors, and fonts)
  - Prerequisite knowledge you can expect from your users
  - Big-picture learning goals of the project
  - Specific and observable learning objectives the users must reach
  - Measurements your client will use to consider whether the training is successful
  - Size of the average learning module
  - Whether users should go through the modules in a predetermined sequence or access individual modules as needed
  - How users will navigate from one module to the next
- **Design and test.** Never assume to know your users' preferences. Most learning is complex cognitive, psychological, and social. It's different for everyone. A sampling of users will help to guide your decisions. The purpose of a good design is to prevent frustrating and wasteful rewrites. So somewhere along the way (hopefully before you start developing), you'll need to stop designing and "freeze" these fundamental decisions:
  - Create a storyboard for one module, proposing a "look-and-feel" for the CBT.
  - Get feedback from target users and buy-in from all stakeholders.
- **Make a paper or computer-generated prototype of each section, representing the navigation elements.**
  - Test the prototype with target users and get feedback from them as well as buy-in from stakeholders.
  - Make a prototype of each section.
  - Once again, get feedback from users and buy-in from stakeholders.

Document the important decisions to protect yourself from flaky clients who may forget what they agreed to.

Creating a Storyboard

For the developer, the storyboard is like the blueprint of a building. It outlines what goes where and the relationship between the pieces. A *storyboard* is basically a hand-drawn or computer-generated sketch of each element of the screen that marries the text and graphics, as you see in Example 16-3.

Storyboarding isn't a glamorous phase of development, because of its repetition and trivialities, but the devil is in the details. It's better to recognize an inconsistency with your pencil poised over a storyboard worksheet than to recognize it with your pencil poised above your final quality assurance (QA) testing. For example, if you wrote a storyboard giving instructions to Zeb (the Martian in Chapter 2), you may notice that you have the image of the jelly jar on the screen, but the voice-over is talking about the peanut butter.

If you're doing the development yourself and the modules are pretty much alike, you might get away with just one "master" storyboard. However, if you're going to delegate or if there's a significant variation between modules, create a storyboard for each module.

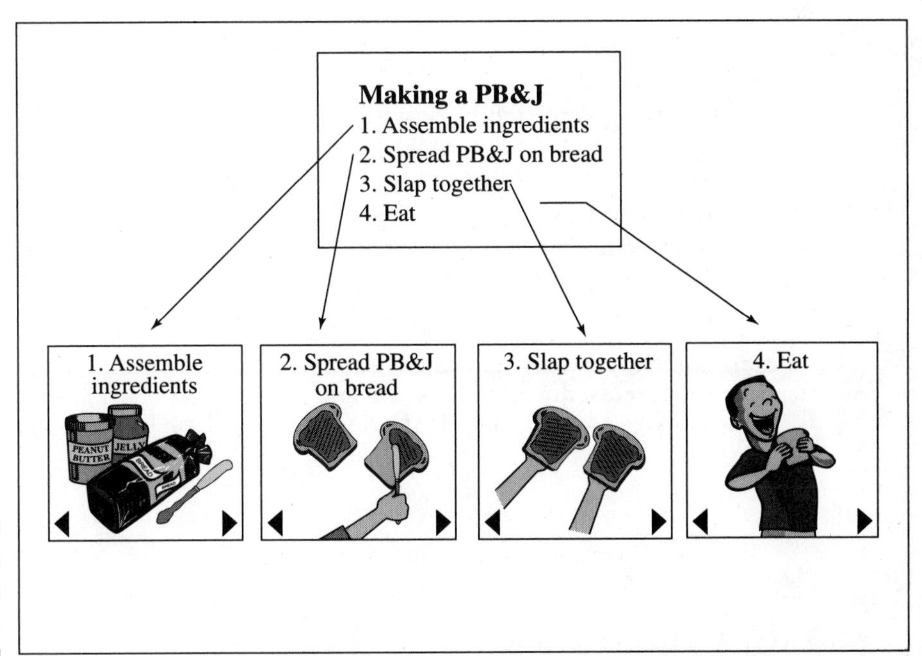

Example 16-3: Sample storyboard.

Planning the Right Kind of CBT

Now that you understand how CBT is disseminated and how it interacts with the users, it's time to plan the CBT that you need to deliver.

Time is ticking

You may be thinking that you can sit back and say, "Now, how much time do I need to do a good job with this?" (Ha!) An unfortunate fact of life is that the average product manager starts to think about training just about the same time he starts to think about taking a vacation after the release date. Therefore, training is generally developed in a rush, with everyone scurrying to meet a nearly impossible ship date.

So the question is rarely, "How much time do I need?" It's "What can I do in the amount of time I have?" If this is your first CBT project, before you estimate how much time you will need, consider creating a prototype. Interactive CBT adds a layer of complexity that translates into more development time. Murphy's Law says that it always takes you at least twice as long as you think.

You need to estimate what needs to be done and design a solution. Here are some milestones in the CBT process:

1. **Design the navigation mechanisms.** These include buttons, menus, interactive images, and any bits of responsive interaction.
2. **Create a sample module and field test it with target users.** Trust me, you need to do this, even if you test it only with a few people on your laptop computer during lunch.
3. **Provide QA on your prototype to be sure that it works on the computer platforms of target users.** It's a real bummer to find out that your CBT looks gorgeous on your own machine but it's slow or scrambled on someone else's. (Don't hoot; this really happens.)

Start doing QA as soon as you have something people can test so that you can fix and retest any problems you find. This process isn't like WBT, where you can make changes and get the document back on the Web in a flash.

Money mania

In addition to time and all your usual expenses for writing and testing, think about your software development tools and the availability of test subjects and test machines. These may be expenses your client hasn't anticipated. Here are some questions to consider:

- Can you get free access to test subjects and computers that replicate your target users' experience?
- Will you have to buy access to either of these important resources?
- Do you have the software tools you need?
- Are there productivity packages that would speed your work?
- If you're including multimedia elements, does your machine have the oomph to handle resource hogs such as video editing and file compression?

Keeping up with maintenance

Whether you or someone else is responsible for maintaining the material you build, account for the maintenance requirements in your design. Schedule periodic upgrades of the CBT to follow the periodic upgrades of the products used to create the components. Following are important questions to ask:

- **How often will the content be updated?** If you're writing training about something that changes every month, you need to make changes that are quick and easy to disseminate. For this type of training, consider delivering on the Web.
- **What material will be updated?** Text is easy to update; video isn't.
- **What's the life expectancy of this training?** If you're going to be living with this baby for the next decade, plan to update it with plain vanilla tools such as a text editor or a stable product from a reputable company. Five years from now, you don't want to be hunting for an old version of some fly-by-night video editing package to update something stored in an ancient file format.

Getting the Goods Out the Door

Think about how you're going to distribute the training. The two key issues are file size and the kind of disks your users can read. Following are yet more questions to ask. (I know this chapter has lots of questions, but you must ask the right questions so you and your client make intelligent decisions.)

- **How big can your files be?** With physical media, the limit is obviously the size of a CD-ROM, which is huge. If you're shipping files over the Internet, the file size depends on the speed of your users' Internet connection and how long they're willing to wait for a download.
- **What media can your typical user read?** Be aware of the variables in equipment. For example, a 100MB drive can't read a 250MB disk. If there isn't one safe medium, plan for alternatives.

Meeting Expectations

Last but not least . . . I can't say enough about how important it is to know your users. Don't settle for just a description of who they are. Try to interview some typical users. You have a better chance of meeting their needs if you understand where they're coming from and what's important to them. For example, you may prepare the smartest content in the world, but if your navigation methods annoy your users, they're going to finish the training feeling annoyed. Or worse yet, they won't finish at all.

When new multimedia tools come out, they're often overused, abused, and inserted into ridiculous places just because it's possible. (It's like a little kid who learns a new word and runs around interjecting it at every opportunity.) Don't succumb! Not only do useless decorations and delays make users grumpy, but they often *date* the CBT. Using these tools is akin to shopping for clothes. When you pick the trendiest style, it's going to be totally out of date next season. If you want your CBT to last as long as your best suit, stick with simple and classic. You know — the charcoal gray or navy blue variety.

Having said that, glossy layout and keen multimedia do have two important results:

- If they're well done, they add significantly to the credibility of your product. Think of them as the voice and attire of your trainer. When people see top-notch layout and graphics, they assume it's top-notch content. Unfortunately, if you hide valuable content under a scruffy cover, your users may not stick with you long enough to discover the gem. You need to know whether users respond better to the pinstriped, silver-haired authority or to the rumpled young nerd in sandals. Then give your CBT the look and feel to meet those expectations.

- If you're training on complex concepts, animation, sound, and/or video quickly convey ideas that would take pages and pages of text.

Testing for Quality Assurance

You need to find and fix the little gotchas and glitches in your CBT. That means you (or some very patient, dogged, dedicated person) must go through every single screen and choose every single menu item and every single button to be sure every single one does every single thing it's expected to do. (As you may have guessed, I stress "every single.") The best way to record errors is to print the screen, circle the errant item, and note the problem. It's a real coup if you can get a member of the target audience to do this, too.

Chapter 17

Web-Based Training: CBT on Steroids

In This Chapter

- Knowing when Web-based training is the right medium
- Using frames and multimedia
- Writing a storyboard and using links
- Using the right tools for the job
- Publishing on the Web

We must not be misled to our own detriment to assume that the untried machine can displace the proved and tried horse.

—John K. Kerr, former major general, U.S. Army

I call this chapter "Web-Based Training: CBT on Steroids" because Web-based training (WBT) is rooted in CBT, but it has more muscle. An advantage of delivering on the Web is that you can update chapters and fix errors with no loss of inventory or distribution time. The updates are completely invisible to your users, and users don't have to wait for you to deliver a new CD-ROM. You don't keep inventory, don't incur shipping charges, and don't accumulate out-of-date copies in your file cabinets.

Because WBT falls under the umbrella of computer-based training (CBT), I suggest that you read Chapter 16 before you read this chapter. It's important to familiarize yourself with some of the commonalties, including types of interactivity, the learning experience, solving business problems, platform independence, designing a process, creating a storyboard, meeting expectations, and testing for quality assurance.

This chapter is somewhat like a primer, explaining how WBT differs from CBT. If you want an in-depth book on creating WBT, check out *Creating Web Pages For Dummies,* 5th Edition by Bud Smith and Arthur Bebak (Hungry Minds, Inc.), and *The Non-Designer's Web Book* by Robin Williams and John Tollett (Addison Wesley Longman, Inc.).

Setting Your Sights (Or Sites)

The decision to create WBT takes a lot of thought. You're ready to make that decision after you and your client answer "yes" to some or all of these questions:

- You want to provide the users with convenient access to the training at any time.
- The content changes periodically, and users always need the most recent version.
- You need to deliver your training to users on a wide range of computer equipment.
- The users have fast, reliable access to the Internet.
- The users are geographically dispersed.
- The users are comfortable downloading plug-ins from other Web sites in order to play your multimedia morsels.
- You have the time, money, tools, and expertise to develop WBT.
- The training will be used by enough people or over a long enough period of time to recoup the investment in design and development.

And you answer "no" to these questions:

- Nanosecond response time is critical.
- You need many large graphics files to display instantly.
- You need a lot of high-quality, full-motion video.
- Content is culturally sensitive with many important nuances.
- Evil forces would like to steal your content.

Planning the Users' Experience

Sit down and give some thought to the purpose of the training experience before you start the actual design. Think about these big-picture questions:

- Are you trying to reduce support calls or increase product quality?
- What are your client's goals that require WBT?

Display the Evidence

When designing for the Web, you have limited control over how the training pages will look on your users' screens. You don't know what kind of machine your users will have or whether your WBT will be displayed on an artist's billboard-size screen, as you see in Example 17-1, or on Granny's old classic, as you see in Example 17-2. Browser software usually does a pretty decent job of making images look good on any machine; however, it's not under your control.

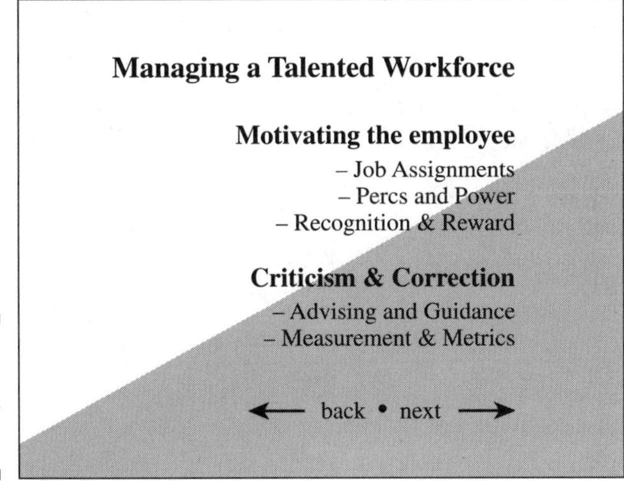

Example 17-1: How it looks on your screen.

Managing a Talented Work
Motivating the em
– Job Assign
– Percs and
– Recognition & R
Criticism & Corr
– Advising and Gu

Example 17-2: How it may look on Granny's screen.

You can direct your WBT to open a window that's 1024 pixels wide so that users will see the layout in the same dimensions that you planned. If a user's computer maxes out at 640 pixels wide, she has to scroll sideways and back and forth to read each 1024-pixel line. Doing so would be incredibly annoying.

Getting chunky

When you write WBT, prepare your information in chunks (also referred to as modules). When you break information into small chunks by topics, it's easy for the reader to absorb the content. Planning your chunk size (and testing samples on target computers) is particularly important when you have large images or multimedia files.

Navigating around

An example of good navigation is a video game in the arcade. You may not know how to use it when you put your quarters in the slot, but within a few seconds you learn what works. The same concept applies to WBT. Following are three rules that are essential in WBT navigation:

1. Provide cues (such as section titles) so users aren't disoriented when traveling in cyberspace.
2. Give users a graceful way to exit.
3. Let users know how to return to a prior screen.

Smooth navigation is important to the success of your training. If users feel confused or disoriented, they won't focus.

Multimedia Madness and Frame-ups

Multimedia and frames add delightful seasoning to WBT when used modestly. *Multimedia* refers to text, graphics, sound, and perhaps movies. *Frames* is a Web-speak term for design elements that divide a Web page into segments — somewhat like a cartoon strip. Each segment contains a separate chunk of text or graphics.

Multimedia madness

Keeping your multimedia under control means knowing how much is within your range. If your client already has finished footage and you only need to clip, compress, and coordinate, then you can probably deliver a product with a significant amount of video. If you need to scavenge or produce footage, that's a different story. You can do it, but it will take longer than you think. Here are a few tips:

- **Name your files and store them in folders.** Keep track of filenames and where they're stored. You do this through naming conventions described in Chapter 18.
- **Keep everything in sync.** Like any manufacturing process, there's an efficiency of scale. With so many phases in preparating multimedia, you may be tempted to say, "I'll just do those last 124 files later because these other 876 are ready to go now." However, if you make the same decision at the next phase and the next, you have mini-collections of files at different stages of development. You'll drive yourself crazy trying to keep track of what's where.

It's a frame-up

Frames can be helpful or send you and users to the nuthouse. You see commercial sites with wildly riotous sets of noncoordinated frames that seem to have nothing to do with each other. (If you're designing training for people who read *Wired* magazine, you can do this.) However, most users want their training presented in an organized fashion, not randomly spattered. And most of the managers want their employees to reach a level of competence, not a level of distraction.

For the Web developer, frames are like rabbits. You start with 2 innocent little rabbits, then you have 12, then you have 100, and then suddenly you can't see the ground because it's hopping with happy little bunnies and you're completely overwhelmed and don't know what to do and someone should have told you that it can get this bad this fast. (If you started to speed up as you read this paragraph, stop and take a breath before you read any further.)

Well, frames are a little like rabbits. One frame is easy to deal with. Two frames can make the user interface much easier for the user. But after that, you have to be very careful. Example 17-3 shows what frames may look like. You notice they're much like frames that cartoonists use when drafting comic strips.

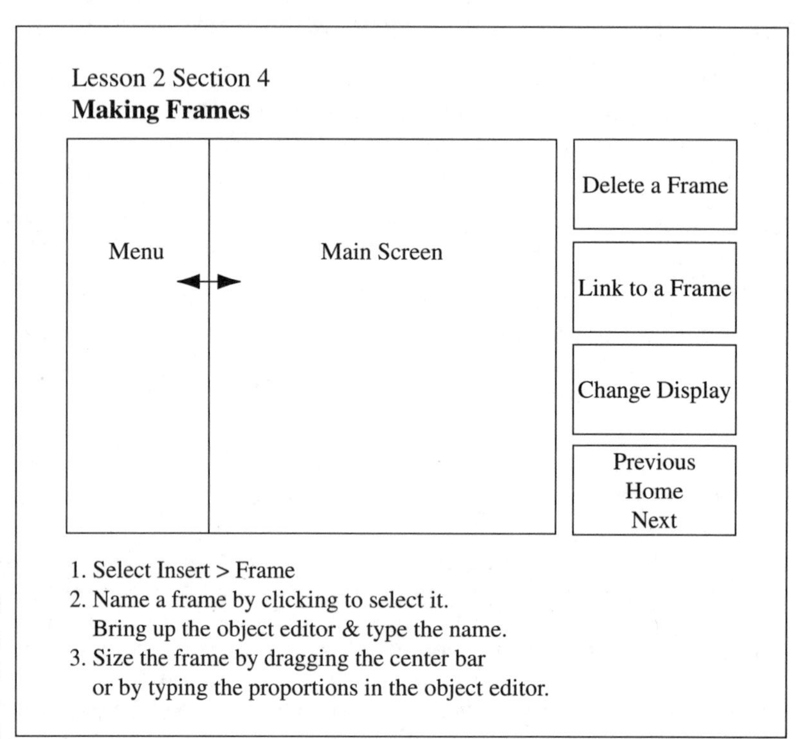

Example 17-3: Becoming a framer.

When you use frames, be very specific with your links. You need to indicate what you want to display and where. Where do you expect the link to display? In a new window? In the originating frame? In the "other" frame of the original window? Example 17-4 shows frames with the text filled in.

Test every link on every page. You must test all the links in all the different frames, which may be enough to make you want to stop writing WBT and apply for early retirement.

Overview of Procedure
Stakeholder List
Phase 1 & Review
Phase 2 & Review
Phase 3 & Review
Finalization
Documentation

Spec Review Procedure
Overview

The basic premise of the Spec Review Procedure is that a wide range of perspectives on a product will increase the likelihood of success of that product. At the very least, it can decrease the likelihood of overlooking something that is very obvious to one group, although it might be unknown to others.

One of the objections to the procedure is that "many cooks spoil the broth". And although it can be argued that operating alone, or in small groups can be faster,

Overview of Procedure
Stakeholder List
Phase 1 & Review
Phase 2 & Review
Phase 3 & Review
Finalization
Documentation

Spec Review Procedure
Overview

The basic premise of the Spec Review Procedure is that a wide range of perspectives on a product will increase the likelihood of

Overview of Procedure
Stakeholder L
Phase 1 & R

Spec Review Procedure
Overview

Previous Section Home Next Section

Example 17-4: Fabulous frame ups.

Writing the Story and Linking It

It's very helpful to write a storyboard that lets you map out how users will move through the training experience. A storyboard is basically a hand-drawn or computer-generated sketch that shows the relationship between the text and graphics. You see an example in Chapter 16.

I go over my own storyboards myriad times, checking every possible option in every possible sequence. For example, if I wrote a storyboard giving instructions for Zeb (the Martian in Chapter 2), I wouldn't want the image of the jelly jar on the screen while the voice-over talks about the peanut butter.

After I'm convinced my storyboard is accurate, I take it to members of my target audience who find glitches. This is a tedious and thankless process, but it's better than having users find the errors in the finished Web page.

Standard Types of Presentations

Here's a list of some common ways to present information on the Web:

- **Text pages and illustrations:** Word processing software makes it fast and easy to develop pages. With nice fonts, a good layout, and fine illustrations, you convey information in a professional and competent manner.
- **White space and landmarks:** Use plenty of white space and give users landmarks (memorable illustrations). Using these techniques gives users a sense of moving through your Web page. They don't like floating through a trackless, featureless space.
- **Sequences and slide shows:** Presenting information as a series of mini-events is an effective training device to build suspense or a gradual understanding. Add a voice-over and you're golden.
- **Annotated illustrations:** Use text annotations to present information centered around a diagram or illustration. Following are a number of variations on this theme:
  - Prepare three frames: one with the image, one with a menu of topics, and one to display the explanation or commentary. Users click on the topic of choice and read the text.
  - Use sound annotations rather than text. Doing so allows users to keep their eyes pinned on the picture.
  - Make the picture the menu so that users click on a link and the annotation is proffered. This is a good way to show more detail. Click on a link and have a window pop up with a detail or cut-away.
- **Compare and contrast:** It's instructive to juxtapose multiple displays. Be careful that you don't lose control of your frames, and heed the text with the Caution icon in the section "It's a frame-up" so users don't get confused about where to look or click.
- **Video:** Video is such a rich medium. Think of how a 15-second video clip can be priceless to show a feature, a subtle or multifaceted activity, or a consequence such as "Here's what happens when you . . ."

Standard Types of Interaction

You may prepare WBT to present concepts or to quiz your users on the material. Sometimes the quiz is for the users' edification to give them an objective evaluation of their competency level. In this case, you don't need to collect their answers.

Keeping score

Sometimes a quiz is mandated, such as for Occupational Safety and Health Administration (OSHA) compliance. When it isn't mandated, the decision whether to save and score is fairly negotiable. I always do my best to talk clients out of scoring because it presents many legal pitfalls. Also, users relate better to training when they don't have Big Brother breathing down their necks.

Choosing the type of questions

Typical quizzes come in a variety of formats: multiple-choice, true-or-false, fill-in-the-blank, essay, and click-and-drag. You may consider mixing them up so you address the needs of varied users. Example 17-5 shows a screen of typical short-answer questions.

If I'm not collecting data, my favorite trick is to make a form on my Web page — perhaps a set of buttons or a text field — and tell users to answer the questions. They commit to an answer by clicking or typing. Write "Click here to check your answer." When they click, the correct answer pops up.

Standard interactions with forms

○ True
◉ False

Fill in the Blank: [ ]

Multiple Choice
○ A
○ B
○ C
◉ None of the above

Short answer: [ ]

Example 17-5: Short answers only.

Tools of the Trade

The process for building WBT depends on which tools you use. You'll probably be working on a tight schedule with a limited budget and will do the job using the tools you already have. Word processors, presentation editors, graphic editors, animation editors, and multimedia editors can all generate Web-worthy material. Every software company is eager to be Internet compatible, so even if your version of the software can't export to the Web, check the latest release. It may be there by now.

Word processors

If you're designing a small training experience that's largely a linear document of text and graphics, you'll do just fine with a regular word processor. For example, the latest version of Microsoft Word has the ability to save documents as Web pages. Because word processors were designed for creating paper documents, they're good at creating Web pages that pretty much scroll down the screen. You can mix and match, bring in pictures and multimedia clips, and link to anything you can dream of.

If you're planning a hierarchy of multiple pages that need consistent titles, fonts, and formatting, you can use the word processor's scripts or templates. Remember, however, that once you enter script land, it's more complicated to go in and tweak the HTML (standard language for documents on the Web) directly because the HTML tags are intertwined with the script tags.

Presentation editors

Presentation editors are optimized for slide shows. They often have sophisticated animation and multimedia capability and are fairly easy to use. However, not all presentation editors reliably export the fancy stuff as Web pages.

To remedy this, you can package presentations in such a way that they're self-contained units, and your Web page can summon those units by simply linking to the file. (Check out "Calling External Programs from Your Web Page," later in this chapter.)

Presentation editors usually include ready-made templates with professionally designed graphics, colors, and fonts. If your publication plans call for clip art, this is a quick way to get a very professional look with no investment in design.

Graphics, animation, and multimedia editors

If you have a keen image, sound, or movie, or if you're a graphics whiz, use a graphics editor, animation editor, or multimedia editors to output Web-compatible files.

However, be aware that the Web-compatible files aren't Web pages. You'll need to "invent" a Web page to launch the files, glue them together, and forge them into a navigable learning experience for your users. The "glue" is HTML — the stuff that Web pages are made of. The HTML can be something as simple as a Web page with a single menu that points to the images or movies. Adobe PhotoShop is one package that automates an elegant solution by giving you forward and back buttons to move through a gallery of images.

Web page editors and Web site editors

Why would you use a Web site editor? For the same reason you drive an SUV in the snow or use a chef's knife to chop onions. A specialized tool does a better job.

Web editors do a lot automatically and give you many options via dialog boxes. Tabs enable you to scoot between the various browsers and editing modes, point-and-click routines let you make hot spots on images, and other options give you the power to change the typeface on titles through the entire site. Web editors are akin to power steering and Smart TV rolled into one.

In addition to the straightforward things you can do by using general tools, Web editors make it easy to do the following:

- **Create original screen layouts.** It's possible to do this with the general tools, but you're going to be wrestling with the software, which is completely convinced that it knows what you *really* want to do, regardless of what you tell it. Web editors make it very easy to change background colors or wallpaper images, or colors of links.
- **Manage the site.** Good Web site editors give a visual map of a site to see the pages that are linked together. This makes it easy for you to keep track of where you are while you're editing, and it avoids the circular loops that confuse users. Site editors fix links when you move files between folders and notify you of trouble spots.

- **Make hot spots in an image.** You could show Zeb (the Martian) a photo of your kitchen counter. (To find out more about Zeb, turn to Chapter 2.) When he clicked on the jar of peanut butter, he'd link to something helpful, such as a definition or graphic.

Some tools tend to favor one browser over the other. So what looks fantastic on one browser may not translate well to other browsers. For example, if you use Microsoft tools, you can include background images in single cells of a table. They display beautifully in Internet Explorer. Netscape Communicator, however, doesn't display a cell's background image.

CBT authoring systems

There are several applications (such as Hot Potatoes from Half-Baked Software) that are optimized for creating interactive training on the Web. These tools automate development of many standard features of online training. Mostly they give you a point-and-click interface for creating Java or JavaScript goodies. You find Hot Potatoes at `http://web.uvic.ca/hrd/halfbaked`.

Some CBT authoring tools (such as Macromedia's Authorware) let you save your training for delivery on CD-ROMs or on the Web. Authorware is incredibly powerful at handling user interactions, but not all that power can be converted to Web pages.

If you have a lot of training to do or will delegate the work to developers, a CBT authoring tool could be a good investment. Do this only if you need standardization or high-octane productivity. The tools are expensive and take time to learn. Consider these applications if your design includes much of the following:

- Branching or feedback that reacts to user input
- Moving objects around the screen in response to users' input
- Changing the display in response to users' input
- Using complex user actions repetitively
- Collecting data

Testing packages

If your clients want to collect data from their users, look into a testing package. These packages have various levels of sophistication, data collection, and data manipulation.

TestPilot gives you a fill-in-the-blank interface to create multimedia quizzes. It integrates with FileMaker Pro, so all the user data is stored in a database. Your client gets a spreadsheet of the test responses with final scores and percents and all those other good data-ish things. Your client, however, will need to put FileMaker Pro on the server.

Dynamic HTML and JavaScript

Dynamic HTML (DHTML), which is a robust HTML, and JavaScript can add a great deal of flash and dash to your Web pages. They are more accessible than most programming languages because you can often find what you need and simply copy data without really understanding much of what those arcane commands do. Many sites on the Web provide libraries of JavaScript and DHTML routines to be copied and adapted by anyone.

You also find a growing number of tools to develop DHTML by pointing and clicking. If you need a lot of give and take between your training modules and your users but don't use an authoring system, you might look into this option. For more detail, see *JavaScript For Dummies,* by Emily Vander Veer (Hungry Minds, Inc.).

Know the users' browsers (and versions) because DHTML doesn't translate well. DHTML is practical for the corporate environment where people generally have the latest of everything. This may not apply to people working on home PCs.

Java

Java is a full-fledged object-oriented programming language, so it's not for the technically faint of heart. If you designed a training experience that includes a sophisticated user interaction (such as a game), then Java is a good tool. Web browsers come with (free) Java run-time engines, so your users have what they need to run your applet (little application) without downloading anything new. For more details about how to incorporate existing Java applets into your material, get a copy of *Java For Dummies,* by Aaron E. Walsh (Hungry Minds, Inc.).

Building a Modest WBT

Okay, you have your design in hand, and you're ready to start the creation process. Always have two versions of backups on two different media. (For example, back up on your hard drive and Zip disk.) Following are some basic tips to help you avoid file-and-folder purgatory.

In the development phase, make a folder for your materials. You want all the links in your pages to be relative, not absolute. That means you want your Web page to look for an image in "the daughter" folder called images and not a folder called C:/images. This is important because, once you move your Web page to the server, all the links pointing inside your folder will work and all the links pointing to your hard drive will break.

Organizing folders

Think about how to organize folders inside your outer folder:

- **Know if there's a lot of repetition by modules.** If so, consider whether to have each module inside its own folder so you can copy and edit the similar files and keep the filenames and links the same, or whether each module will reuse the exact same files. Keep a folder just for these multiple-use files. The bigger your project, the more important this is. Just imagine doing this 201 times.
- **Put images into separate folders.** If you have a gazillion images, you may want to have separate folders for icons, photos, and animations. If you have more than a few dozen files in a folder, they're cumbersome to find.
- **Keep your sources in a separate folder.** When you finally publish to the server, you won't waste server space with unused files and won't comb through each folder removing the sources.

Naming files intelligently

Naming files intelligently becomes more important as your Web pages get bigger and more complex. Even with a modest-sized Web page, it's wise to start out with intelligent names. Don't name your files pbj-1.html, pbj-2.html, and so on. Instead, call them overview.html, jars.html, or spread.html.

Also, if you'll be moving back and forth between Macs and PCs, limit your filenames to eight characters, using only letters, numbers, and the underscore. More elaborate filenames may get garbled when you move from one operating system to the other.

Get all your images and multimedia materials into their proper folders before you start linking to them. A good site editor can manage these moves for you, but preplanning is critical. If you make links to your images and then move them to another folder, the links will break because they'll be looking for the images in the old location.

Now that you have your folder structure set and your files in their final folders, you can start bringing the pieces together by following the blueprint of your storyboard. Write your text, insert your files, make your links, and save your Web page.

Calling External Programs from Your Web Page

Following are multiple reasons to go outside the bounds of your Web page and give your users other applications or applets:

- Someone wrote something relevant that's not a Web page. You want to provide it to your users.
- You want to give your users a program that's easy to develop with other tools (versus developing with a Web development tool to run through the Web browser).
- You need to have fast response time, so you want an application to run locally on your user's computer (not over the network).

Documenting and Archiving Sources

Web browsers display only a few types of graphics files, so chances are that you're going to create your images with one file format (such as PhotoShop format) and then save it as something that a Web browser can display. GIFs and JPEGs are both compressed file formats. JPEGs are ever-so-slightly fuzzy versions of the original. If you're saving as JPEG files, save the original in case you need to alter it later.

GIF and JPEG files are representations of a multicolored grid of pixels. The fancier graphic editors (such as Macromedia Fireworks) store more intelligent representations of an image, letting you label and layer objects.

Quality Assurance Testing

Take a leadership role in testing your WBT. (You may need to be pushy, insistent, and even obnoxious to get this done.) Expect your clients to be generous and forgiving *before* the release date. They may even suggest that you don't need to test because they trust you. Of course they do or they wouldn't have hired you. However, if something doesn't work *after* the release date, they'll be justifiably upset.

When the development phase is complete, test every link and every menu item on every page. You only need to do this once because if the links work on one platform, they'll work on another.

While you're designing, pick a set of platforms you're going to test on. You'll need buy-in from your client and may need to be crafty about getting information. Here's a typical conversation:

> **You:** So, what configurations do we need to run on?
>
> **Client:** Just the regular ones.
>
> **You:** Both PCs and Macs?
>
> **Client:** I'm not sure. Just PCs I guess.
>
> **You:** So we'd better plan for both. Both Navigator and Explorer?
>
> **Client:** Well, I use Navigator. So does Pat. So, just Navigator, I guess.
>
> **You:** Maybe we'd better plan for both. Is everyone running the latest version of operating systems and browsers?
>
> **Client:** Oh, sure. We get upgraded once a year.
>
> **You:** Do you think it's safe to say that no one is running with software more than a year old?
>
> **Client:** Well, I think some people might want to do this at home, and I'm not sure how often they upgrade.
>
> **You:** Is it safe to say that no one is running with software more than two years old?
>
> **Client:** Oh, sure, that should cover everyone. Well . . . except Chris, who uses a really, really old machine when he works at home because his kids are always on the new one.
>
> **You:** Do you think it's okay if we tell Chris to bump the kids or do this work in the office?
>
> **Client:** That'll probably work.
>
> **You:** Okay, I'll guarantee that the WBT will work on two-year-old browsers, two-year-old operating systems, and two-year-old machines. Anybody who has older equipment will need to upgrade or use another system for this WBT.

Publishing to the Web

Because this chapter is about WBT delivered in real time across the Internet (and not shipped and installed on the users' computers), you're not concerned with packaging your material into a kit for shipment. Therefore, this section focuses on how to get your training onto a Web server so that your users can reach it.

Large corporations typically are connected to the Internet via T1 (or greater capacity OC series) lines. To distribute throughout a worldwide corporation, publish the final product to the Web for local users. Then copy to a public server on the company's intranet.

The systems engineers in other geographic locations log on to the (originating) public server and download the application or Web module to a local server. The users always download from the local copy or connect to the application on the server. A master schedule is created, and communications mechanisms (e-mail) are put in place regarding where and when to get modifications.

Getting it off your desktop

The basic idea is that you want to move the Web page folder from your desktop into a specified folder on the client's Web server. Conceptually, it's very simple. You log into the area on the Web server and copy your folder. Here are a few things you need to do this:

- File transfer software, such as FTP or Fetch
- The server name for the client's server
- A username and password so that you can access the server
- The name of the folder in which you put your Web page folder

The owner of the server area may offer to move your Web page folder to the client's server for you because she doesn't want to give out the password or she doesn't trust anyone but herself. Depending on how the server is set up, deleting other people's stuff or putting the material in the wrong place can be easy.

If you're not clear about every step of this procedure, ask for help. My colleague had to put out an all-points bulletin to a Webmaster last year because someone accidentally replaced the corporate Web page with his new product sheet. Apparently he hit the up arrow a few extra times and didn't notice where he was. Incidents such as this do happen!

Controlling access

If you want to control access to your Web page content for security reasons, privacy issues, or commercial reasons, you'll need to password-protect your material. Most control mechanisms limit access to one folder and its subfolders. There are many ways to do this, and nearly all require some kind of system privileges. So talk to your Webmaster and find out what the scoop is for your Web server.

If you want to limit access to local users on your client's intranet, you may be able to do this yourself, but you'll still need to talk to the Webmaster to get the implementation details of that particular setup.

Chapter 18

Creating Online Help

In This Chapter

- Knowing the types of online Help systems
- Establishing a methodology
- Understanding what the users need to know
- Developing the content
- Testing the help document
- Moving a print document online

Men might as well project a voyage to the moon as attempt to employ steam navigation against the stormy North Atlantic Ocean.

—Dr. Dionysus Lardner (professor of natural philosophy and astronomy, University College in London), addressing the British Association for the Advancement of Science, 1838

People work in the community of online help for a variety of reasons. Some opt for careers in writing online documents because of the challenges this evolving technology offers. Others enter the field after writing technical documents in the paper world; they have to make the switch or risk becoming dinosaurs with stale, one-dimensional skills. Still others are part of a technical team who are told that they must "get something online."

This chapter offers a glimpse into the world of online help. I use the word "glimpse" because online help is a very robust media. Learning the ins and out takes a book, not a chapter. If you write online help, check the Web using the search words *online help authoring* for a wealth of information. (Chapter 14 gives a wealth of information about how to conduct a search.)

Getting Intimate with Online Help

The purpose of online help is to give users instant answers to questions or problems and to put the answers where users can find them quickly and easily. However, this isn't as simple as it sounds.

Using special software

In order to write online help, you must become intimate with the software to create it. You don't create online help with your word processing software. Instead, you use one of the popular authoring tools such as RoboHelp, Doc-to-Help, Forehelp, HelpBreeze, FrameMaker, Help Magician, and others.

You can download trial versions of some of these programs from the Web by using the software name as the search word. To learn more about each authoring tool, check the software's online help. Also, look for courses given in your area.

Knowing the types of online help

Online help comes in many flavors — all of which are intended to make the users' journey through the document easier. Online help isn't just an option on your screen with a pull-down menu; it may be anything from an electronic user manual to a tutorial, wizard, agent or coach, tip, pop-up, or context-sensitive help. Following is a brief description of each:

User manual

An online user manual gives users instant access to information without their having to look for a paper manual (that's never around when they need it). Some online manuals have been converted from print documents, known as legacy documents. Later in this chapter you can find a discussion about making this conversion. Chapter 8 offers a wealth of information on creating user manuals.

Tutorial

A tutorial is a step-by-step program that shows how to use features of a program — from simple to complex. Users generally find tutorials helpful because they learn at their own pace, receive feedback, and get to practice tasks by using steps that are relevant.

Wizard

A wizard is a software program that bears a slight resemblance to a tutorial. The key difference is that a tutorial takes the user down a path in a predetermined order. A wizard lets the user select the tasks.

Agent and coach

An agent or coach monitors what users are doing and gives advice (somewhat like your mother did when you were a kid). One example is the dancing paper clip in new versions of MS Office that pops up when you're in the middle of something. The clip carries a message that asks if you really know what you're doing. You may want to shoot the messenger, but the dancing clip is an agent.

Tip of the day

A tip is a brief message that users see when they execute a program. For example, you may tell users about command accelerators or hints for using little-known parts of the program.

Pop-up

A pop-up is a special link that's great for an on-the-spot definition or quick explanation. Pop-ups appear on the screen and overlay what they explain. You see an example of a pop-up in action in Example 18-1.

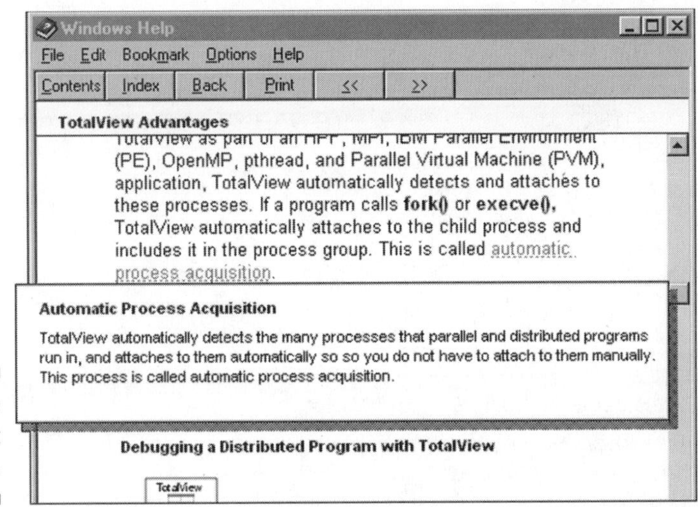

Example 18-1: See what pops up.

Becoming chummy with the programmer

You'll undoubtedly be working closely with one or more programmers (also called software developers or program engineers). The programmer develops the software; you, the author, explain how to use it. Meet with the programmer at the beginning of the project to establish a good working relationship.

Barry, a colleague of mine, was working in Massachusetts and was writing a help system for software written in Germany. Barry spent many weeks on the phone with the programmer in Germany trying to work through a variety of issues. Finally, the client sent Barry to Germany so he could work directly with the programmer.

When the two met face to face, they worked out all the issues and had the system up and running in about half a day. The cooperative give-and-take they established in just one morning made it very easy for them to jointly solve problems that surfaced later. In this high-tech society, we often ignore the importance of face-to-face contact.

Context-sensitive help

Context-sensitive help gives users additional information on a topic when they click on a button or hyperlink. This generally brings up another screen, as opposed to a pop-up. For example, if a user needs help on starting a process, he may go to the help screen that shows the process in detail. In Example 18-2 you see what such a screen may look like.

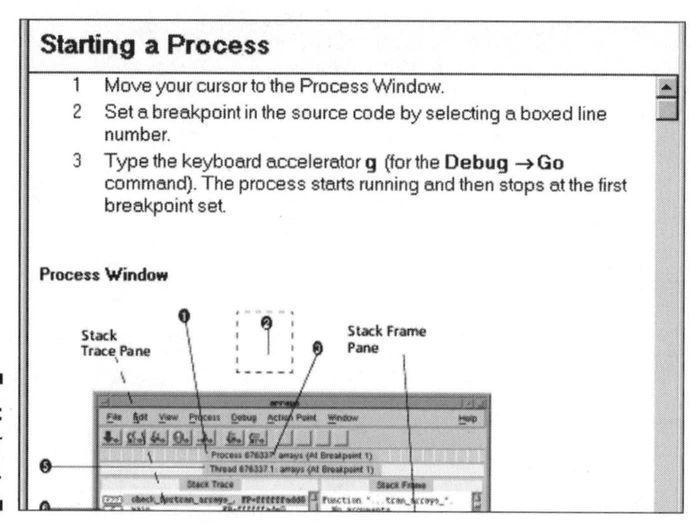

Example 18-2: Help at your fingertips.

Delivering the goods

Online help is delivered in one of three ways: on a CD-ROM as part of an online user manual, as a separate "book," or downloaded from the Web.

Getting Started

Before you write online help, fill out the Technical Brief detailed in Chapter 2 and on the Cheat Sheet in the front of this book. It will help you to determine the best way to deliver the document. For example, when you determine that newbies may have to install the software, consider putting the installation portion of the document in print and the remainder online.

Helping users to get going

In an ideal world, you could write a snappy user manual that gets new users up and running with no fuss, no muss. Since this isn't a perfect world, however, there generally is both fuss and muss. Therefore, you probably need to include an online help section called "Getting Started." Here's what you may include in this section:

- A tutorial or wizard
- A description of the program or application
- Things users need to do before they use the application
- What actions users perform to get the results they want
- The meaning of the results they get

Trying this on for size

To give you some practice in writing online help, assume that you're writing the help for Microsoft Internet Explorer or Netscape Communicator. Here are some questions you may want to ask:

- Do users know what the Internet is?
- Do they have one of the latest browsers?
- If so, would they know how to connect?
- Would they understand how to use a URL?

If you read through Chapter 2 — and I certainly hope you did — you probably remember Zeb from the planet Zeblonia who needed to make a sandwich. Take a glance at the questions you needed to ask in order to help Zeb.

Establishing a methodology

You need to establish a methodology for writing online help — a way to make the authoring process simple. Table 18-1 suggests steps for creating this process:

Table 18-1 Process for Developing a Methodology

Step	*Action*	*Explanation*
1	Define	Define the production cycle from start to rollout. Define your users with the help of the Technical Brief. (Also check with the marketing, sales, human resources, information systems, and other departments who may help to define the users.)
2	Prototype	Understand the philosophy of the design. Identify filenames, hyperlinks, graphics, and more.
3	Plan	Create templates and style guides. Design a file system and file structure.
4	Develop	Develop and incorporate graphics, voice, video, animation, and music, if appropriate.
5	Test	Test on different computers, platforms, and operating systems if your help runs on the Web or is embedded in the application. Revise and retest.
6	Publish	Go through a peer review before you roll out the document. Let everyone know how to find your document.
7	Maintain	Ask for feedback. Once your site goes live, keep the information on the site current. Evaluate your successes and failures.

Striking a Balance

Once you identify target users and address their needs, then you can turn your attention to the disparate groups. You must provide adequate information so you don't bore sophisticated users or frustrate those who are unsophisticated. You can do this through a combination of paper and electronic documents, pop-up screens, and context-sensitive help.

To take this one step further, there are many issues to consider — some of which are more complex than you may think. For example, following are some considerations for writing online help for financial software for disparate users:

- **How do I . . . ?** Your reader may ask, "How do I figure out a monthly loan payment?" It's your job, as the author, to read into this question that the user is actually asking, "How do I compute the monthly payment for a $10,000 loan at 8 percent if I pay back the loan in four years?" Because users don't often ask the right "how do I. . . ." questions, you must anticipate them. Following are some issues as they apply to this question:
  1. You need to describe the procedure to obtain a monthly payment.
  2. You need to figure out a way for the user to find this information. For example, Lotus often has an index in its help menu called "How Do I." After users understand that procedures are located in this area, they start browsing through it. Doing so helps them find the information they're interested in as well as remember other things stashed there.
  3. Because users don't know what users don't know, your job is to make this information available. For example, something as simple as figuring out a monthly payment requires that they know the amount they're borrowing, the interest rate, and the term of the loan.
  4. You never do things in isolation. In almost all cases, you need to link different procedures together so that users see other options.·
- **Where do I . . . ?** As you suspect by now, help systems are big, unwieldy things containing different kinds of information. And, unlike a printed book, this information has no linear order. You must make it clear to users where they can find the information they need.

Ultimately, the best way to learn about help systems and the things in them is to become an "A" student. Look tirelessly at different help systems for the software you use. Evaluate the strengths and weaknesses.

Naming Conventions

A simple online project may require hundreds of files; a complex project may require thousands. One way to simplify this process is to establish standard naming conventions and to enforce these standards vehemently. Following are some of the conventions to establish for each project:

- Topic name
- ID
- Filename
- Author
- Date created/modified
- Keywords (see what follows)

In your naming convention, use a keyword that describes each topic. For example, if you have topics named *Coke* and *Pepsi,* they may not have the word *soda.* You can create an ID to let users retrieve *Coke* and *Pepsi* when they type the keyword *soda.*

Naming conventions are crucial because when the help system finds the topic ID, it displays the associated topic on the screen for the user. Therefore, every topic must have an ID. Be sure to work this process out with the programmer beforehand. Following are three ways to generate IDs:

1. As the author of online help, you create the IDs and share them with the programmer.
2. The programmer creates the IDs and shares them with you.
3. The help authoring tool creates the IDs.

The Litmus Test

Writing and publishing online documents isn't as simple as writing and publishing paper documents. You must run alpha and beta tests before you publish. The alpha test is done by people in house; the beta test is often done by potential customers who agree to be test sites. Table 18-2 shows what to test, where to test, and why to test.

Table 18-2 Testing for the Real World

What	*Where*	*Why*
Locality	• Your campus or building • All the time zones in your country • Other countries	Everything may work well in your location, but it may not work well across the country or across the world.
Browsers	• Microsoft Internet Explorer • Netscape Communicator	Browsers handle text and graphics differently, so you must test Web documents on the popular browsers.
Platforms	• PC • Mac • UNIX • Linux	Platforms handle text and graphics differently. What looks good on one may not look good on another.
Monitor resolution	• 640 X 480 • 800 X 600 • 1024 X 768 • 1280 X 1025	Many users hang on to old monitors, and their lower resolution may distort text, graphics, or animations. Check out Chapter 15 for tips on creating sights and sounds.

Be sure to check all the hyperlinks. Make sure that they work and lead to the correct site.

Software development is always work in progress; therefore, online help is always work in progress because it must reflect all the updates. This is akin to maintaining a garden. Someone needs to pull out the dead weeds and replant the new crop in order for the garden to flourish. In the electronic world, flourishing involves adding enhancements, fixing bugs, and releasing new or interim versions. If the company hires you to "maintain the garden," it's pretty much your job. If not, you turn this task over to the company "gardener."

Moving a Print Document Online

As mentioned earlier in this chapter, one aspect of online help may involve moving print (legacy) documents online. As people become comfortable with electronic media, this is happening more often. Following are some ways to simplify the process so you don't pull your hair out.

Developing the content

Experts agree that people have a more relaxed attitude when reading print than when reading electronic media. With electronic media, users have a mouse in the palm of their hand, and with the quick click, they move on.

Following are some broad brush strokes to capture the users' attention so they refrain from clicking and moving on. For more details, check out Chapter 5.

- **Limit the paragraph and sentence length.** Limit paragraphs to no more than 8 lines. (Yes, that's lines, not sentences.) Keep sentences to no more than 25 words.
- **Riddle with lists.** Use more bulleted and numbered lists than you would use in print because they're eye-stoppers on a screen.
- **Write straightforward headlines and subheadlines.** Doing so helps users skim to find key information.
- **Use hyperlinks.** The authoring tools provide a great way to cross-reference and give users additional information.
- **Write in modules.** Keep in mind that users access help to find answers. This means writing lots of modular text, rather than weaving themes through a large number of pages.
- **Use graphics strategically.** Graphics increase the size of the file. Use them judiciously to clarify concepts, display processes, or otherwise convey information that's better expressed visually.
- **Because space is often limited, use abbreviations or acronyms.** Be certain, however, that the user knows the abbreviation or acronym. If in doubt, consider including context-sensitive help or a pop-up window.
- **Tone down the vocabulary.** Keep the language simple, yet appropriate. Remember that a wide range of people will read the document — including people who may speak English as a second language. (Check out Chapter 6.)

Take one step at a time

The best way to minimize conversion problems is to convert one chapter at a time, test it, and see what feedback you get. Based on this early feedback, you can make changes to the way you write and format the remainder of the document.

Create a style guide

If you're responsible for converting a legacy document, create a style guide so you don't experience a scene such as this: Kim uses pop-ups and jumps to define terms that users may not understand. James places these terms in a glossary. Kim places buttons across the top of the screen. James places them across the bottom. Kim is partial to purple and selects that as her background color. James is partial to green and selects green as his background.

A style guide for an online document goes beyond a style guide for a print document. Both identify layout, abbreviations, font selections, and the like. Online documents also identify the following:

- Placement of buttons (across the top, bottom, or side)
- Background and text colors
- Graphics, logos, and animations
- Video clips
- Naming conventions

Map out the site

Sometimes your online document is part of a larger site that may contain product information, training information, and the like. It's important to have a site plan to see where the online document fits into the big picture. Example 18-3 shows a site plan sketch for a company's Web site. The online documents flow from the library.

Keep On Trucking

Things are very hectic the first time you release a help document. The grim reality is that there's never enough time to do everything right, let alone time to complete everything you want to do. Your goal is to do the best you can within your time constraints and be able to add information later on.

Don't be discouraged if the very first user finds the worst typo or the lamest sentence you ever wrote. It's all part of the process. Smile, learn from your mistakes, and get a life!

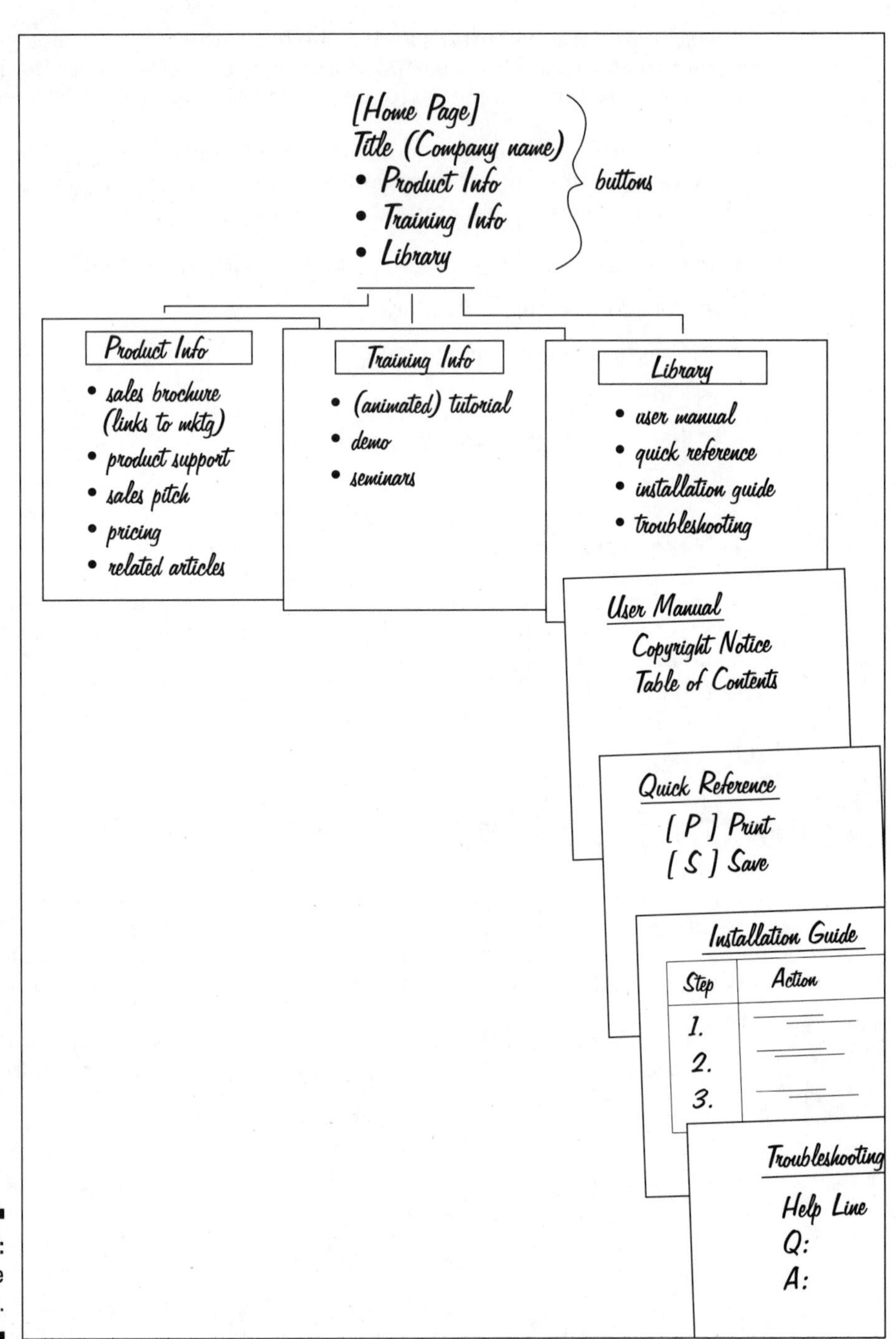

Example 18-3: The whole enchilada.

Part V
The Part of Tens

In this part . . .

No *For Dummies* book is complete without a few ragamuffin chapters. This part — which is reserved for those chapters that just don't seem to fit anywhere else in the book — shows you how to

- Get published in a technical journal
- File a patent
- Write a grant proposal
- Make your technical documents shout, "Read me!"

Chapter 19

Ten Ways to Make Your Technical Documents Shout "Read Me!"

In This Chapter

- Getting to know your reader(s)
- Creating an outline
- Writing a draft
- Striving for visual impact and the right tone
- Realizing how you look when you don't proofread
- Getting input from others
- Working in the electronic realm

The computer does not substitute for judgment any more than a pencil substitutes for literacy.

—Robert McNamara, American politician and statesman

Whenever you pick up a technical document that someone else wrote, notice what you like and don't like about it. Let each document be your teacher. Following are guidelines to make you the poster child of technical documents.

Before you begin to work on any technical document, be sure to fill out the Technical Brief you see in Chapter 2 and on the Cheat Sheet in the front of this book. It walks you through the steps to understand your readers, their level of competence, the key issues, the project team, and executional considerations. Filling out the Technical Brief is a critical step in planning your document.

"See" Your Target So You Know Where to Aim

Reading requires a minimum of two people — a reader and a writer. (And you already know which one you are.) After you use the Technical Brief to understand the needs of your readers, envision them as real people. If you don't know your readers or can't envision them, you can't write to them. Take a moment to close your eyes and ask yourself these questions:

- What do my readers look like?
- What do they do for fun?
- What kinds of cars do they drive?

When you envision your readers, put smiles on their faces. They're not a pack of growling faultfinders out to get you. Unless you tell your readers that the Internal Revenue Service wants to see them immediately, they're on your side and they want to read what you write. After all, you're sharing information of value.

Create Structure with Bones

No human being stands straight without a skeleton — the bones that provide structure. Writing also needs structure. Technical writing's structure comes in the form of an outline. The outline may be the traditional one you learned in school (with roman numerals, uppercase and lowercase letters, and all those indentations), or it may be an annotated table of contents, such as the one you see at the beginning of this book. Whichever method works for you, you must provide a framework from which to begin writing. Chapter 3 has a great discussion of creating structure for your document.

Add Meat to the Bones

After you have a solid outline, writing the draft is like adding meat to the bones or filling in the blanks. As you learn in Chapter 4, start with one section — the one that's easiest for you to write. Your readers will never know where you started. Then proceed to the second easiest, and so forth. Create a comfortable environment, get all your stuff together, set attainable goals, and stay focused.

Don't procrastinate by saying that you have to rearrange your sock drawer or take your cat in for a whisker-ectomy. If you can't get started immediately, try freewriting. *Freewriting* is a warm-up exercise for your mind, just as stretching is a warm-up exercise for joggers. Sit at your computer and start writing . . . anything. Eventually, you start to come around to your topic. Some of my best ideas often come from freewriting.

Make the Bones Visually Appealing

You must give your document visual appeal or no one will read it. Chapter 5 offers a host of tips to create visual impact. Here are a few highlights:

- Create headings and subheadings so that readers can find information at a glance. These serve as mileposts and help readers skim a document for valuable information just as they do a newspaper.
- Use bulleted and numbered lists, when appropriate.
- Limit paragraphs to no more than 8 lines and sentences to no more than 25 words.
- Include tables, figures, charts, and photos, when appropriate.
- Use color judiciously.
- Include navigational aids in your electronic documents. Web-savvy readers expect to find these elements.

Hone the Tone

Certain types of documents often dictate the tone of the writing. Stereotypically speaking, for example, a directory is boooooring. An academic paper is pompous. Advertising copy is hype. Technical documents are mind boggling. Prove the expectations wrong for technical documents: Write documents so they're clear, concise, and geared for the level of your reader. When your writing isn't clear, it's because your thoughts aren't clear. Check out Chapter 6 for ways to hone the tone of your document. Here are a few highlights:

- Say what you need to say in as few words as possible.
- Create a positive tone with positive words. (That may be as simple as stating what to do, rather than what not to do.)
- Use the active voice to bring life to your documents.
- Choose words that are appropriate for your readers.
- Use humor and jargon sparingly (if at all).

Even if you determine that your readers don't have the intelligence to pass an inkblot test, you must still use a tone that shows respect.

Proofread 'Til Your Eyes Pop Out

Chapter 7 goes into lots of detail about proofreading. It's important to proofread until your eyes pop out so you don't wind up with errors such as the ones that follow. The writers of these remarks were either numskulls or poor proofreaders, but in either case, they lost their credibility. Proofread carefully so that you don't lose your credibility. The magic is in the minutiae.

- Charles Darwin was a naturalist who wrote the Organ of the Spices.
- It is a well-known fact that a deceased body warps the mind.
- For snakebite: Bleed the wound and rape the victim in a blanket for shock.
- To collect sulphur, hold a deacon over a flame in a test tube.
- Three kinds of blood vessels are arteries, veins, and caterpillars.
- The thermometer is an instrument for raising temperance.
- Heredity means that if your grandfather didn't have children, then your father probably didn't have any. So you probably won't have any.
- Algebra was the wife of Euclid.
- To be a nurse, you must be absolutely sterile.
- Artificial insemination is when the farmer does it to the cow, not the bull.

Give the Document the Litmus Test

Always get a second opinion. When you write a short document, ask a co-worker or some other reader to give it a sanity check. Is the message clear? Is the information well organized? For a large document, go through a peer review after the first draft is complete. This is the time to get everyone's input on the general content, formatting, and any other issues. If you can, assemble all the reviewers in the same room. If not, gather as many as you can.

Two heads may be better than one, but two egos are worse. Typically, everyone who reviews a document feels compelled to comment. This compulsion is just human nature and doesn't necessarily reflect the quality of your work. Never take comments from others personally, or you may wind up selling pencils on the street corner.

For E-Docs Only

Readers can't flip through electronic documents as they can through paper documents. Following are ways to make your electronic documents shout "Read me!"

Make the home page revealing

Prepare the home page so that when readers view it they know what the document is about and where to find information. (This may be as simple as creating hyperlinks or buttons.) In technical terms, this is called "before the scroll." That's a term derived from the newspaper language "above the fold," which is what you see when the newspaper is folded and sitting on the rack.

Create summaries

In addition to hyperlinks and buttons, consider writing brief summaries — a hyperlink followed by a line or two explaining what the readers may expect to find. This is similar to a descriptive abstract that gives the highlights of an article in a sentence or two. (For the scoop on abstracts, check out Chapter 9.)

Create links that are easy to follow

When you write an electronic document, minimize the number of links the readers must follow. You don't want readers paging through too many links to get to the "pot of gold," and you don't want them missing valuable information along the way.

When you label links clearly and specifically, you tell readers what they'll find so that they can quickly decide whether to follow the link. Notice how a key piece of information turns the weak link that follows into an informative link.

Weak link: Conclusions

Informative link: Conclusions: Omica has a 95 percent success rate

Chapter 20

Ten Tips for Publishing in a Technical Journal

In This Chapter

- Finding the right journal
- Showing your determination
- Understanding the masthead
- Writing the query letter
- Understanding the lingo
- Avoiding the last writes

The concept [reliable overnight delivery service] is interesting and well formed, but in order to earn better than a C, the idea must be feasible.

—Yale professor in response to Fred Smith's paper (Smith went on to start Federal Express)

At some point in your career you may have "pearls of information" to share. The way to reach the masses is by publishing an article in a journal or other professional publication. Whether the publication appears in print, on a high-profile Web site, or in both, the potential readership is enormous.

People who publish are part of an elite group. And the more prestigious the publication, the more the value! Not all publications pay for articles or offer honoraria; the payoff is in getting published. When you publish an article, you *add prestige to your reputation and your company's or institution's reputation.* You can order reprints to include in sales packets and proposals. And the best part is that you're not marketing yourself or your product; the publication is doing it for you!

A reprint of an article you write is a great promotional piece to attach to your resume. After you have something appear in print, you're considered an expert.

Getting *write* to it

Before you start to write your article, use the Technical Brief in Chapter 2 and on the Cheat Sheet in the front of this book to understand your audience and key issues. Write your article by using the guidelines in Parts I and II of this book.

If your topic is very revolutionary or controversial, you may face some obstacles in getting it published. Editors aren't necessarily crusaders. Crusaders in the publishing industry either receive journalism awards or wind up with their golden futures behind them.

Don't Procrastinate; Just Do It!

If you're intimidated thinking about the prospect of trying to get something published, don't be! Most editors are clamoring for good material. If you ever had responsibility for publishing a newsletter, you know how difficult it is to get contributors to submit interesting articles. Editors have the same problem.

Even if you're a newbie and have never been published, don't let that stop you if you have something worthwhile to contribute. Remember that Mark Twain, John Grisham, Mary Higgins Clark, and Sheryl Lindsell-Roberts started somewhere! (I'm certainly not in their league, but I can dream.)

Don't underestimate the value of determination. Before I got my first article published, I must have sent it to every publisher in the universe. I had gotten so many rejection letters, I could have wallpapered the Taj Mahal with them. However, I never gave up and am now a professional writer with a slew of articles and 18 books to my credit. If I did it, you can too!

The famous photos of President John F. Kennedy's assassination were submitted by an amateur photographer — one who never had anything published. The photos and the photographer received worldwide acclaim.

Hooking Up with the Right Publication

You're a professional. You know what publications you and your colleagues read. If you're unsure where to submit your topic, contact professional organizations and ask for suggestions. Or contact the publication and ask for demographic information about readers.

Before you contact the publisher, however, read at least a half dozen back issues of the journal you target. Doing so will give you a good idea of the type of articles the journal looks for, the style and tone to use, the length of the articles, and lots of other useful information.

Lurking Behind the Masthead

The masthead is the front part of the publication that lists publishers, editors, phone numbers of branch offices, board of reviewing editors, member societies, and other good stuff. The only ones who read mastheads are the mothers of the people mentioned and wannabe writers (who read them insatiably).

At first glance, you notice that nearly everyone listed is an editor of some sort. Trying to decipher who does what is the real challenge. The best way to find the editor who accepts unsolicited manuscripts is to call the publication or check the *Writer's Market* — an annual publication that tells you who published what.

Understanding the Lingo

Following are a few phrases that you may see in the *Writer's Market* (or similar publications) and what they really mean:

- **"We don't accept unsolicited manuscripts."** Gee whiz — doesn't this sound like the publication wants to discourage freelance submissions? Absolutely not! Editors depend on submissions to keep the publication afloat. They simply want to discourage the amateur who hasn't taken the time to learn the ropes.
- **"Reports promptly."** Yes, they may report the rejections promptly. They pass around queries that are being considered.
- **"Reports in two to four weeks."** Two to four months may be more like it.
- **"Pays five to ten cents a word."** You can bet you'll get five.

Writing a Query Letter

You may compare a query letter to a cover letter that you send with a resume. It's the first thing the editor reads. The query letter gives a clear indication of your writing style and thought process. The difference between a cover letter and query letter is that the former accompanies your resume; the latter stands alone. It's not wise to send your article unless the publication requests it. Therefore, the query letter either piques or squelches the editor's interest.

What to include

Limit the query to one page and be sure you include the title of your article. If the title is tentative, refer to it as a "working title." Here are some things to focus on:

- Start with a hook — an attention grabber. For example, does the article present anything new or insightful?
- Stress how you intend to approach and develop the topic.
- Explain what photographs or other graphics you have to support your data.
- Give your best estimate of how long the article will be (in approximate number of words).
- Specify why you're qualified to write the article. You may attach a resume or other data to support your qualifications.
- End with a request to write the article.
- Specifically ask the editor to respond.

Proofread until your eyes hurt. And include a self-addressed, stamped envelope, known as a SASE, to simplify the response.

Follow up

If you don't get a response within four weeks, it's appropriate to give the editor a call. He's probably buried under a deluge of other queries and may not have read yours. (You may not connect with an editor, but it's worth a try.) Once you pique the interest of the editor, he may present your idea to a review board. You may not be notified this is happening.

Following is the body of a query I sent to *Northeast Sailing Life* that resulted in getting my article published. I opened with a hook, gave the gist of the text, indicated the length, and concluded with a call to action.

Dear [name]:

A favorite conversation among sailors has always been boat names. Each boat owner likes to share "his story." Therefore, I thought Northeast Sailing Life would be an ideal publication for boat owners to do just that.

Title: "What's in a Name: It's Absolute Lunar Sea"

I sent out questionnaires to dozens of sailboat owners and have accumulated dozens of funny and heartwarming stories. "What's In a Name? It's Absolute Lunar Sea" would make wonderful reading for your sailing audience. Please let me know if you'd be interested in seeing my 1500-word manuscript with a view toward publishing it. In addition to stories, many boat owners sent photographs that would be ideal to use.

I look forward to hearing from you and sharing these wonderful stories with your readers.

Sincerely,

Sheryl Lindsell-Roberts

Simultaneous Submissions

When a publication requests a copy of your manuscript with a view toward publishing it, the publication likes to feel that it has it exclusively. *So don't send your manuscript to more than one publication at a time.* If more than one publisher wants to read your manuscript, put your priorities in order and send the manuscript to one publication at a time.

If you don't hear within a month, it's appropriate to call the editor who requested the manuscript and ask for an update. If you don't get a decision within two months, it's then appropriate to contact the editor and mention that you'd like to know of his interest because another editor is waiting in the wings. (Say that only if it's true.)

Confidentiality

Writers are often concerned about the confidentiality of a manuscript. When you deal with a reputable publication (and you know who the big names are), you *can* trust the editor to hold your query in strict confidence. He won't share it with competing publications or disclose your ideas for building a better mousetrap. Therefore, give him whatever information adds strength to your query.

Be very careful, however, about including patentable ideas, trade secrets, financial information, or anything else of a confidential nature. Before you do that, check with an attorney or an officer of your company.

Writing a white paper

In some countries, a white paper is an official government report detailing government policy that's to be voted on by the country's legislature. In the technical and marketing arena, however, a white paper is somewhat like the un-cola. It's not quite a report; it's not quite a marketing piece; and it's not quite a sales piece.

A *white paper* is an informational paper that gives readers unbiased information on a topic. You write a white paper and get it published in much the same way as you do an article. A white paper may be published in a journal, periodical, or on a Web site. When you write a white paper, you and your company get credit for the writing, although your product (or service) isn't mentioned in the text. The payoff is that your name and company appear as having produced the paper, so it's free advertising!

Following is the opening of a white paper for a company specializing in product-specific training. Check out the neutrality of the tone.

> *To get an immediate return on your Customer Relationship Management (CRM) investment, you must get your employees up and running immediately. Companies can invest millions of dollars in state-of-the-art, front-office enterprise applications to try to keep customers for life. Once the companies make the investment, then what? Their employees sit in front of their computers, scratch their heads, wrinkle their brows, and wonder how to use the application.*
>
> *In the absence of knowledgeable, confident, and well-trained employees, a CRM enterprise application rollout is money down the drain. Although some CRM vendors claim to weave training into the cost of the application, the effort is often little more than one-time, one-size-fits-all training. This is totally ineffective because one size can never fit all.*

The white paper goes on to discuss challenges that companies face, how to get an immediate return on investment (ROI), and what to look for when selecting a training company. The white paper is informational, yet slanted toward the training provided by the company for whom it was written.

Don't Take "No" for an Answer

Some publications readily accept unsolicited manuscripts; others "tell you" they rely on staff writers. (I put *tell you* in quotes for a reason.) Every publication worth its salt knows that input from a broad range of contributors strengthens its publication. Even if the editor tells you that he doesn't accept unsolicited material, don't let that dissuade you from sending a query. You'll be hard to ignore if your topic is impressive and presented exquisitely.

If one publication rejects you, keep sending queries to others. A rejection isn't necessarily a reflection on your topic or your writing. It may merely mean that the topic isn't appropriate for the particular publication or the publication doesn't have space available at the present time.

Chapter 21

Ten Things to Know about Filing a Patent

In This Chapter

- Understanding what constitutes a patent
- Submitting your invention
- Getting professional advice
- Filing a patent in a foreign country

Everything that can be invented has been invented.

—Charles H. Duell, former commissioner of the United States Patent Office (1899), when he wanted to close the patent office

In 1995, a woman named Jeanne Calment celebrated the 20th anniversary of her 100th birthday (in other words, her 120th birthday). During her lifetime, she witnessed the invention of antibiotics, automobiles, biological weapons, bombs, computers, inoculations, machine guns, poison gas, space travel, television, and a whole lot more. In her 120 years, Ms. Calment's life probably changed more often than Elizabeth Taylor's last name. So, Mr. Duell, how wrong were you?

Anyone can apply for a patent regardless of age or mental competency. Even deceased people can file a patent through a personal representative.

Each year, thousands of people add prestige to their names and the names of their companies and universities because of inventions. If you invent something, you want to exclude others from making, using, or selling it. Patents, therefore, are intended to promote innovation and make sure that inventors reap the financial rewards for their inventions. When you're granted a patent and someone violates it, you can sue for patent infringement. Once a patent expires, however, anyone can make, use, or sell the invention or its design.

Each country has different regulations for filing patents. The information in this chapter relates to the United States only. For information on filing patents in other countries, check with the country's patent office or its equivalent.

Types of Patents

Several types of patents exist. The two most common are utility and design patents.

Utility patents

Utility patents cover inventions that are electrical, mechanical, or chemical. These may include staplers, electronic circuits, pharmaceutical products, microwave ovens, semiconductor manufacturing processes, and much more. They may even include genetically engineered bacteria for cleaning up oil spills. Utility patents expire after 20 years.

In order for an invention to be granted a utility patent, it must meet the following specifications:

- **Its use must be obvious.** Does the invention produce new or unexpected results?
- **It must be novel.** Is there some novel feature, or is the invention merely a new use for something already done?
- **It must fall into a statutory class.** That includes processes, machines, articles of manufacture, compositions, or new uses of another item.

Did you know that since the U.S. Patent Office opened its doors in 1838, 4,400 mousetraps have been patented? Who ever said, "You can't build a better mousetrap"?

Design patents

Design patents cover inventions that have a unique shape. This spans everything from computer screens, to the design of a water cooler, to telephones shaped like Elvis's guitar. Design patents expire after 14 years.

With some design patents, however . . . you just *gotta* wonder. Example 21-1 is a baby carriage invented by George Clark. It looks like a high-buttoned shoe with wheels underneath. Mr. Clark invented this carriage so that wealthy people could push their spoiled brats around in a carriage no one else had.

Example 21-1: Keeping kids in shoes.

And who said you can't reinvent the wheel? That's exactly what Sydney Jones of Great Malvern, England, did. He invented a wheel made of elastic spring steel. Its rims fold around obstacles and roll over them, somewhat like an air cushion, as you see in Example 21-2.

Webs and dots

A lot of controversy surrounds Web patents. A patent holder obviously gains exclusive dominance of a technology and prevents its competitors from vying in the same market space. That creates a monopoly by the controlling company. (Microsoft's dominance is currently being tested in the courts, and it will be interesting to see how this ultimately shakes out.)

Additionally, the new dot-com companies are getting patents for words they use on their Web sites. For example, Amazon.com holds the patent for *One Click* technology, and Priceline.com holds the patent for *name-your-own-price* as a way to sell goods over the Web. Companies such as these are winning lawsuits when these patented names are violated. For example, BarnesandNoble.com is barred from using *one-click* checkout. Its site uses two clicks to check out.

Example 21-2: Reinventing the wheel.

What's an Invention?

The answer to this question isn't that simple. Major controversies are growing over the difference between *inventions* and *discoveries*. *ComputerUser* editor James Mathewson wrote, "We don't invent new mathematical truths, we discover them." For example, as an outgrowth of patenting hybrid crops, biotech companies are filing patents for genetic sequences. This development has many scientists displeased. They feel that one day a person or company may own the rights to complex animals or humans.

If two or more people work together on an invention, in whose name is the patent? If they are joint inventors, the patent is issued to them jointly. However, if one person provided all the ideas and one or more merely followed instructions, the person who conceived the ideas is listed as the sole inventor.

He Who Hesitates Is Lost

Perhaps Alexander Graham Bell got tired of his kids sending him telegrams asking for money, so he invented the telephone just to screen his kids' calls with call waiting. (Just kidding!) Even the telephone, however, created a controversy. Elisha Gray supposedly submitted a preliminary patent application to the U.S. Patent Office for a telephone. Later the same day, Bell submitted a full patent application. *Ma Gray* was the loser.

What is intellectual property?

You can patent intellectual property (IP), which refers to a product of the intellect or mind. This may include a business method, software, formula, or industrial process. IP spans principles that determine who owns intellectual property, who may be excluded from exploiting the property, and the degree to which the courts are willing to enforce the owner's rights.

Doing Your Homework

The U.S. Patent Office doesn't require that you make a search to see whether your invention has already been patented; however, why put effort into an invention if someone has already claimed it? There are several ways to make your search:

- **Check out stores, catalogs, product directories, and more.** For example, if you invent a newfangled paper clip, check large stationery outlets and catalogs to see whether something like it is already on the market.
- **Check the Web and do a search by using the words "patent search."** You find more sites that you know what to do with. For more information on conducting a search, check out Chapter 14.
- **Visit the U.S. Patent Office.** Why not enjoy the sights in and around the nation's capital and have a wonderful vacation or business trip?

CTEA and the Magic Kingdom

A patent differs from a copyright. A copyright protects authors, composers, artists, and others against anyone from having their work copied without their permission.

The framers of the Constitution established 14 years as the term of a copyright, with an additional 14 years if the author is still alive. With people living longer, terms were extended. Today, for a work-for-hire piece, the term is a maximum of 95 years; for an individual, it's the life of the author plus 70 years.

Until 1998, the term of a copyright was 50 years after the death of the author. However, the folks at Disney Studios shuddered at the thought that the copyright on Mickey Mouse would run out, so they made significant donations to Congress. Alas, the Sonny Bono Copyright Term Extension Act (CTEA) keeps Mickey out of the public domain for a while longer.

Submitting Your Idea

You don't need a finished product to apply for a patent. Before you get into the nuts and bolts (literally), prepare a detailed sketch of your idea. Example 21-3 shows a patent drawing submitted by Jon A. Roberts, Karl J. Armstrong, and Arnold J. Aronson for a planarization method used in the semiconductor industry. In addition to the sketch, here's what the inventors included as part of the packet:

- Abstract
- Background of the invention
- Summary of the invention
- Description of the drawings
- Detailed description of the invention
- Supplementary drawings

What to send

Send a black-and-white drawing, generally done in India ink or equivalent that makes solid black lines. The U.S. Patent Office accepts color drawings for utility patents, but only after you file a petition explaining why color drawings are necessary.

The patent office doesn't generally accept photographs for utility patent applications. If a color photo is important to displaying your idea, file a petition requesting that a color photo be accepted. If you send a photograph, develop it on double-weight photographic paper or permanently mount it on bristol board. The photos must be of sufficient quality so that all details in the drawing can be reproduced in the printed patent.

Submit ample sketches to show as many views as necessary to detail your invention.

Go with the flow

If your invention is software related, prepare a flowchart showing the process. Check out Chapter 5 for more about flowcharts. Make sure that your flowchart is complete enough for any skilled programmer to follow.

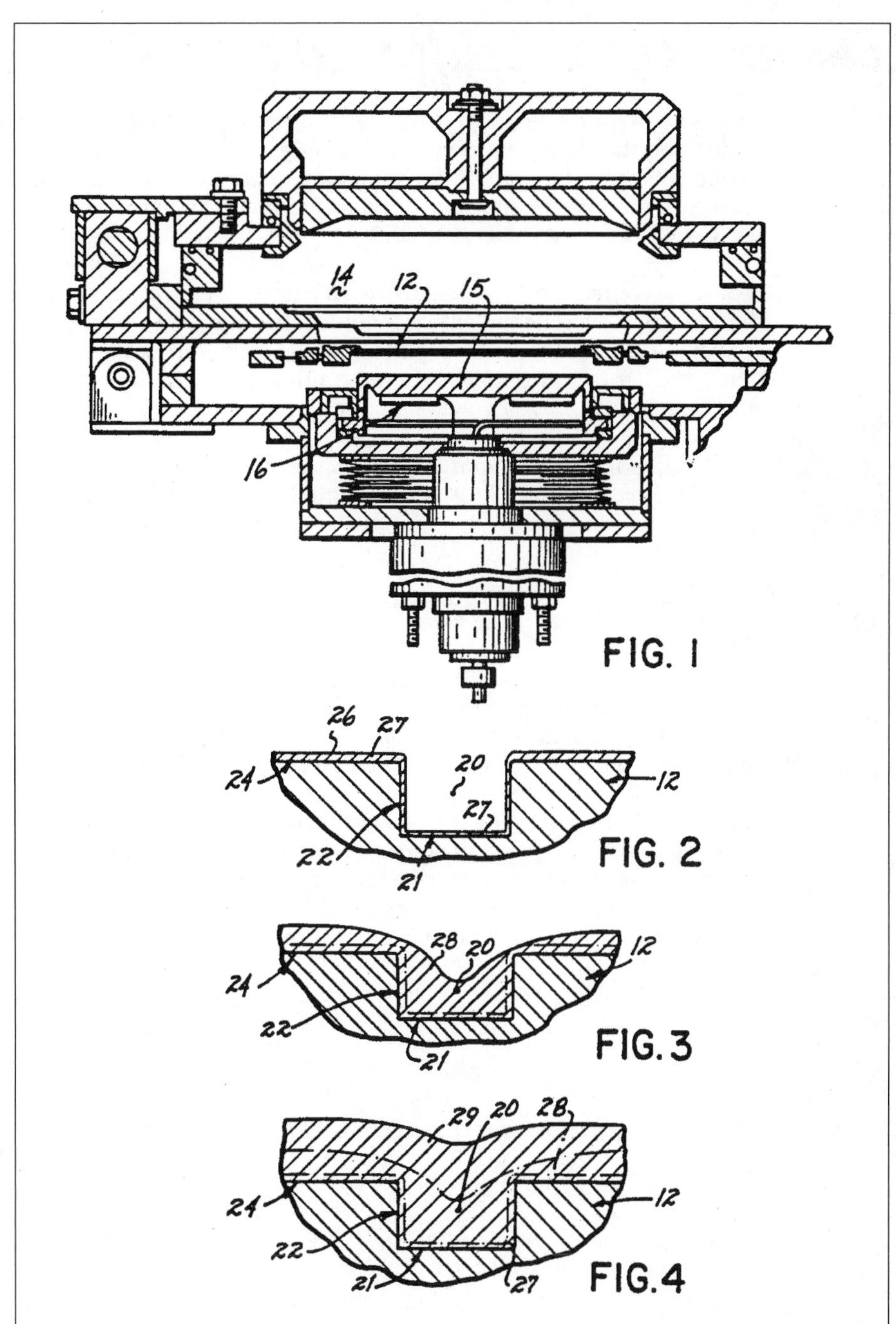

Example 21-3: Drawing submitted for patent approval.

Leaving Legal Stuff to the Pros

If you work for a company, get advice from the company's attorneys. They'll guide you through the process and handle all the filings. Although your name will be on the patent, the company generally holds the patent. (When you endorsed your employment contract, you may have signed a clause to that effect.)

For do-it-yourselfers, it's wise to hire a patent agent or patent attorney. A patent agent is someone with a technical background who can conduct a patent search and is authorized to voice an opinion. However, a patent agent can't appear in court should there be a problem with her findings. It's wise to hire a patent attorney who can handle the whole nine yards.

Get a copy of *Attorneys and Agents Registered to Practice Before the U.S. Patent and Trademark Office* (A&ARTP) at a medium- to large-sized library.

On the Foreign Front

U.S. patent laws have no effect in a foreign country. Therefore, a patent granted by the United States extends only throughout the U.S. and its territories. Countries have their own patent laws, and you must make an application for a patent in the country where you want your invention sold and protected.

The Paris Convention for the Protection of Industrial Property is a treaty relating to patents that 140 countries adhere to. It provides that each country guarantee to the citizens of other countries the same rights in patents it gives to its own citizens. If you contact a U.S. patent attorney, she may be able to set you on the course of acquiring a patent attorney abroad.

Chapter 22

Ten Tips for Writing a Grant Proposal

In This Chapter

- Knowing what to include in a grant proposal
- Learning from your experience

They laughed at Columbus. They laughed at the Wright Brothers. But they also laughed at Bozo the Clown.

— Carl Sagan, astronomer

Grant money is an avenue open to research groups and the nonprofit sector to get funding for worthwhile projects. For instance, you may be part of a university team that's on the brink of a breakthrough in preventing or treating cancer and needs money to continue the research. At times like these, you may be called upon to request funding from a donor.

Donors are waiting in line to fund worthwhile projects; they may be federal and state agencies, community foundations, international organizations, and others. If you don't have a specific donor in mind, check out the Internet. Conduct searches for words such as *foundations, grants,* or *funding.* You'll be amazed at what's out there. (Chapter 14 is chock-full of ways to do a search.) This chapter provides you with some general tips on writing grants. For more detailed information, see *Grant Writing For Dummies,* by Beverly Browning (Hungry Minds, Inc.).

The process of asking for grant money is embedded in the belief that a partnership between the soliciting organization and the donor will result in a dynamic collaboration. Your job is to convince the intended donor of that. Base your request on the idea that you have the capacity to solve a specific problem or need but don't have dollars to implement the program.

The extent of information you include is relative to how large the project is and how much money you request. For large proposals of more than ten pages, include any or all of the following sections. For smaller proposals, you may omit the executive summary, appendixes, and glossary.

- Title page
- Table of contents
- Executive summary
- Introduction
- Statement of need
- Project description
- Budget description
- Conclusion
- Appendixes
- Glossary (if terminology isn't familiar to reviewer)

Include with your proposal any brochures, testimonial letters, or other data that may strengthen your request.

Title (Cover) Page

The title page includes the title of the project; the name, address, and telephone number of the project director; the requester's name; and the project's beginning and ending dates. Following are some suggestions for making the title page eye-catching:

- Prepare the title page on high-quality paper that's thicker than the paper used for the inside pages. This paper is often called "card stock" and comes in a variety of colors and thicknesses.
- Consider using one or two colors for the text. You can do a small proposal in your word processor and print it with an inkjet printer. (Check out Chapter 5 for information on colors that are appropriate for different purposes.)
- For a large proposal requesting large sums of money, consider having a graphic artist prepare the title page.

Table of Contents

If the proposal is ten pages or longer, include a table of contents so that the grant reviewers can find what they need at a glance. Consider the following suggestions:

- Include subheads when you think the reviewers may need breakdowns of information.
- Use leaders (....) to connect the text and the page numbers. Notice how effectively that's done in the Table of Contents in the front of this book.

Executive Summary

The purpose of an executive summary is to provide the reader with a high-level snapshot of what's in the proposal. (Chapter 13 gives a wealth of information on writing a dynamic executive summary.) Keep in mind that you're probably not writing for a favorable or neutral audience. This audience needs to be convinced. Here's what you may include in the executive summary for a grant proposal:

- Brief statement of the problem or need
- Short description of the project
- Total funding requirements

Introduction

In order to present a winning proposal, you need to provide the information you see in the list that follows. If you're submitting a short proposal requesting a few thousand dollars, this may be all you need for the proposal to be complete.

- **Concept:** Start with how the project fits into the philosophy and mission of your facility. Articulate the concepts well and make them compelling. Donors want to know that a project reinforces the overall direction of an organization.

- **Program:** Here's some of the information you'll need to round up:
  - The nature of the project and how it will be conducted
  - The timetable from start to finish
  - The anticipated outcomes and how best to evaluate the results
  - Staffing needs (existing staff and new hires)
- **Financials:** You probably won't be able to pin down all the expenses until the details and timing have been finalized. At this stage, you sketch out the broad outlines of the budget to be sure that the costs are in proportion to the outcomes you anticipate. If it appears that the costs will be prohibitive, scale back your plans or adjust them to remove expenses that aren't cost-effective.

Statement of Need

After you successfully pique the interest of the prospective donor, your next job is to justify the need. Present the facts and establish that you understand the problems and can reasonably address them. If possible, include testimonials from authorities in the field as well as from your agency's own experience.

The following is from a grant proposal to continue research for an early detection process for esophageal cancer. Notice how the writer presents the need for this research.

Esophageal cancer is the ninth most common cancer in the world. According to the American Cancer Society, only 12 percent of all white patients and 8 percent of all African-American patients survived after 5 years of diagnosis of esophageal cancer in 1992, but the median survival time is 9 months. Although the survival rates have been improving, most people with esophageal cancer die because of advanced diagnosis, at which point it is likely for the cancer to have metastasized. With an early detection system, we can greatly increase the survival rate by an anticipated 50 percent.

Project Description

Be succinct, yet persuasive. Like a good debater, assemble all the arguments. Present them in a way that will convince the reader of their importance. As you marshal your arguments, consider the following:

- **Highlight the facts or statistics that best support the project.** Be sure that the data are accurate.
- **Be optimistic and enthusiastic.** For example, if you're looking for funding for breast cancer, you may say something like this:

  We know that breast cancer kills. But statistics prove that regular checkups catch most breast cancer in the early stages, thereby saving lives. Therefore, a program to encourage checkups will reduce the risk of death for women diagnosed with the disease.

- **Propose this program as a model by explaining how your solution can be a solution for others.** This approach can expand the base of potential funding. Make this argument only if it fits.
- **Toot your own horn.** Does your program address the need differently or better than other projects that preceded it? Describe your program without being critical of the competition. The potential donor may have invested in these other projects.
- **Keep in mind that today's donors are very interested in collaboration.** They may even ask why you're not collaborating with those you view as key competitors. So, at the very least, describe how your work complements the work of others without duplicating it.

To describe the project, include its objectives, methods, staffing/administration, and evaluation.

Grants for lulus

In my book *Goofy Government Grants & Wacky Waste* (Sterling Publishing, Inc.), I reveal some lulus that were funded — many of them with our tax dollars. I don't know the outcome of these studies, but I could have learned a lot from reading the proposals.

- The National Aeronautics and Space Administration (NASA) spent $200,000 to see whether sweet potatoes can be grown in space.
- The National Institute on Alcohol Abuse and Alcoholism spent $102,000 to discover whether sunfish that drink tequila are more aggressive than sunfish that drink gin.
- The U.S. Department of Agriculture (USDA) wanted to see whether pregnant pigs would be less stressed if they jogged. So they devised treadmills for the pigs at an undisclosed sum of money.
- The National Science Foundation spent $229,460 to study the sexual habits of houseflies.
- The National Endowment for the Arts funded a program for a woman named LeAnn Wilchusky to board a small airplane armed with a large bundle of crepe-paper streamers she threw into the sky. This space sculpting cost us $6,025.

And if you think the United States has a monopoly on funding wasteful projects, the Japanese government funded a seven-year program to determine whether catfish cause earthquakes when they wiggle their tails.

Objectives

Clearly state your objectives. They must be measurable, tangible, specific, concrete, and achievable in a specified time period.

Clear: Our after-school remedial education program will help 50 second-grade students improve their reading scores by one grade level. This will be measured in standardized reading tests administered after they've participated in the program for six months.

Vague: Our after-school program will help children read better in a short time.

Methods

Describe the specific activities that will take place to achieve the objectives. It may be helpful to divide the methods into how, when, and why.

- **How?** Give a detailed description of how the project will run from beginning to end. Your methods should match the objectives.

  CytoPath will take this existing technology to modify and retrain the system's software so that it can work effectively to identify atypical cells in the esophagus. Esophageal cancer is easy to treat in its early states.

  By using the CytoPath system, a clear and detailed view of various types of cells are obtained. With this technology, the cells are scanned through a microscope that takes pictures of the cells that are most likely to be abnormal. If there are abnormalities detected, the findings are examined by a cytotechnologist.

- **When?** Present a timetable that tells the reader all the pertinent details of the timing issues. Include the start date, milestones, and end date.
- **Why?** Present your methods, especially if they are new or unorthodox. Why will the planned work lead to the outcomes you anticipate?

  The CytoPath technology has been found to detect a false negative diagnosis in 92 percent of tested women about one year prior to the diagnosis by manual screening.

Staffing/Administration

Include the staffing details here or in an appendix. The placement of this information depends on its length and importance. For example, if your staff includes people whose names are recognized in the industry, include them here. If not, you may put the names and biographies in the appendix.

- Describe how you plan to staff and administer the project. Include volunteers, consultants, and permanent staff members. (Include salaries in the budget section.)
- Devote a few sentences to discussing the number of staff members, their qualifications, and specific assignments.

Evaluation

An evaluation can often be the best means for you and others to learn from your experience. There are two ways to proceed with a formal evaluation:

- Measure the product.
- Analyze the process.

Either or both may be appropriate to your project; the approach you choose depends on the nature of the project and its objectives. For either, highlight the manner in which evaluation information will be collected and how the data will be analyzed.

Be very careful about including patentable ideas, trade secrets, financial information, or anything else of a confidential nature. If this information is necessary to convey the understanding of the project, check with an attorney or an officer of your company or institution. And get their permission in writing in order to CYA (you know, cover your anatomy).

Budget

The budget may be as simple as a one-page statement of projected expenses or an extensive presentation. Make sure that your presentation is as complete as it can be so there won't be surprises later on. You may want to divide the expenses into the following areas:

- Personnel
- Travel
- Equipment
- Printing
- Testing
- Anything else you can foresee

Conclusion

Every proposal should have a concluding paragraph or two. Here are some tips for wrapping it up:

- If appropriate, outline some of the follow-up activities you may undertake to prepare your donors for your next request.
- State how the project may become self-sustaining and carry on without further grant support.

Appendixes

Typical appendixes may include biographical sketches of key people involved in the project, letters of support, statistical tables, cost documentation for equipment, audited financial statements, and anything else that supports your request. If your proposal includes terms that the grant reviewer may not understand, follow the appendixes with a glossary.

Don't forget to use your Editing Checklist found in Chapter 7. If you send out a proposal with errors, the grant reviewers may construe that as a sign of the way you do your work. Your grant can be a large goose egg.

Sweating it out

Waiting is perhaps the most tedious part of the grant writing process because the ball's no longer in your court. If your hard work results in a grant, acknowledge the donor's support with a letter of thanks. If not, there's always next year.

If your proposal for funding is rejected, contact the donors (or in this case the un-donors) and find out why. You want to learn from the experience. Ask these questions:

- Did they feel your request lacked merit?
- Did the donors need additional information?
- Would they be interested in considering the proposal at a future date?

If the grant reviewers didn't fund you for the latter of these reasons or for any other reason you can remedy, put them on your mailing list so that they can become further acquainted with your organization. Maybe the next time you'll have better luck. If not, remember the words of the great W. C. Fields: "If at first you don't succeed, try, try again. Then quit. There's no use being a damn fool about it."

Appendix A

Punctuation Made Easy

Punctuation is one of the most significant tools you have to create documents that bear the mark of your own voice. When you speak aloud, you constantly punctuate sentences with your voice and body language. And when you write, you make a sound in the reader's head. Your "writing voice" can be a dull, sleep-inducing mumble (like a tedious, unformatted document) or it can be a joyful sound, a shy whisper, a throb of passion. It all depends on the punctuation you use.

I present the punctuation marks in the order in which they're most commonly used and confused.

General guidelines

Here are a few general punctuation guidelines:

- **Place commas and periods inside quotation marks.**

  The engineer wanted to hear her supervisor say "yes."

- **Place semicolons and colons outside quotation marks.**

  The design used is "Taguchi L16c"; it consists of 16 wafers.

- **Place question marks and exclamation points inside the quotes only when they apply to the quoted material.**

  "Did you read Sally's findings?" Jim asked.

- **Place question marks and exclamation points outside the quotes when they apply to the entire sentence.**

  Did the supervisor say, "Set the pressure to 10 psi"?

You can also use punctuation to stress what you want your readers to see as important. For example, the following three sentences are worded identically. Yet the different marks of punctuation give each a unique sound:

Dashes: The Ace Chemical Company — winner of the annual Service Award — just introduced its new product line. (The dashes emphasize the award.)

Parentheses: The Ace Chemical Company (winner of the annual Service Award) just introduced its new product line. (The parentheses downplay the award and emphasize the introduction of a new product line.)

Commas: The Ace Chemical Company, winner of the annual Service Award, just introduced its new product line. (The commas neutralize the entire sentence.)

Commas

Commas are the most frequently used (and misused) punctuation mark. While periods indicate a *stop* in thought, commas act as *slow signs* — like speed bumps. They let you know which items are grouped together, what's critical to the meaning of the sentence, and more — as the general rules listed below indicate.

- **Use commas to separate three or more items in a series.** A comma before the final *and* or *or* is optional. It can, however, increase clarity. The choice is yours, but be consistent.

  We'll need cartridges, staples, and copy paper.

- **Use a comma before a conjunction (*and, but, or, nor, for, so,* or *yet*) that joins what could be two complete sentences.**

  Joining two sentences: Joe recognized the four presenters, but he couldn't recall their names.

  One sentence plus a clause: Joe recognized the four presenters but couldn't recall their names.

- **Don't place a comma before *because.***
- **Use commas to separate items in an address or a date.** But don't use any punctuation before a zip code.

  On Monday, April 2, XXXX, Cranston Technology Co. will move to 13 James Street, Maxinkuckee, IN 46511.

- **Use commas to set off an expression that explains or modifies the preceding word, name, or phrase.**

  Jim Smith, our IT specialist, will be on vacation the week of May 3.

- **Use commas to set off one or more words that directly address the person to whom you're speaking by name, title, or relationship.**

  Please let me know, Marv, if you can add anything to those findings.

- **Use a comma after an introductory phrase if it's followed by a complete sentence. This type of clause may include introductory words such as *when, if, as,* and *although.***

  If we state our case clearly, we should get the funding.

- **Use commas around a phrase that isn't necessary to the meaning of the sentence.**

  Barbara, whom you met at the office last week, is a speaker at the semiconductor conference. (Barbara is a speaker at the conference regardless of when you met her.)

- **Don't place commas around information that makes the sentence clear.**

  The person who meets all our qualifications will never be found.

- **Use commas to set off expressions that interrupt the natural flow of the sentence.** These expressions include *as a result, in fact, therefore, however, consequently, for example, in fact, on the contrary,* and others.

  We will, therefore, continue with the project.

- **Use commas to clarify a sentence that would otherwise be confusing.**

  It may be a long, long time before we get the test results.

  In 1999, 53 computers were sold. (Or: In 1999, fifty-three computers were sold.)

  Without Bill, Jim can't proceed.

  Only three weeks before, I had lunch with him.

- **Use commas for emphasis.**

  The shipment, unfortunately, was delayed.

- **Use commas to show contrast.**

  The assignment is long, but not difficult.

- **Use commas to identify a person who is quoted directly.**

  John Naisbitt, American business writer and social researcher, said, "We are drowning in information but starved for knowledge."

- **Use commas to set off designations, titles, and degrees that follow a name.**

  Ted Adler, President of Verdox Company, will be next month's speaker.

- **Use a comma to divide a sentence that starts as a statement and ends as a question.**

  I can't think of anything further, can you?

- **Use commas to separate items in reference material.**

  You can find the Peano-Gosper Curve in *The Fractal Geometry of Nature,* by Benoit B. Mandelbrot, Chapter 7, page 70.

- **Use a comma to separate words when the word *and* is omitted.**

  Please include a stamped, self-addressed envelope.

One space, not two

In the olden days of typewriters, spacing twice after a punctuation mark was sound advice. After all, doing so was the only way to clearly separate one sentence from another. With computers, however, we have proportional spacing that shows a clear distinction between sentences.

Therefore, you should *space once* after a period, colon, exclamation point, question mark, quotation mark, or any other mark of punctuation that ends a sentence.

Colons and Semicolons

Semicolons are separators that are stronger than commas and weaker than periods. Colons direct the reader's attention to what follows. The following sections show how to use them correctly.

Semicolons

Consider semicolons a cross between periods and commas. They create more pause than commas, yet less than periods. Following are some rules about when to use them:

- **Use a semicolon in place of a conjunction *(and, but, or, nor, for, so,* or *yet)* to join complete sentences.**

  The Georgia plant supplies the raw material; the Chicago plant provides the finished product.

- **Use a semicolon when a parenthetical word (*however, therefore,* and the like) or phrase introduces a separate sentence.**

  The project came to a standstill during the strike; however, we did eke out a small profit.

- **Use a semicolon to separate items in a series when the items themselves have commas.**

  The milestones were January 15, 2000; February 20, 2000; and April 15, 2000.

Colons

Colons are marks of anticipation. They serve as introductions and alert you to a close connection between what comes before and after it.

- **Use a colon after an introduction that includes or implies *the following* or *as follows*.**

  These are the people you will meet: James Smith, Jerry Alexander, and Bob Nethers.

- **Use a colon to introduce a long quotation.**

  Professor Longwinded said: "The project came to a standstill after . . ."

- **Use a colon to separate hours and minutes.**

  He should arrive at 10:45 a.m.

Dashes and Parentheses

Dashes and parentheses affect how readers understand information. Dashes highlight the text; parentheses play down the text.

Dashes

Dashes (often considered strong parentheses) are vigorous and versatile. They can stand alone or be used in pairs. Just don't overdo dashes, or they lose their impact.

Here are two ways to form an em dash: If you don't space between the second hyphen and the word that follows, the two hyphens magically become an em dash. Or in MS Word you can look in the pull-down menu Insert and then highlight Symbols. Some writers leave one space before and after the em dash, others leave no spaces. Check your company's style guide. If there isn't one, pick the one you think looks best.

- **Use dashes to set off expressions you want to emphasize.**

  This application — as unbiased tests have disclosed — is more powerful than what you're currently using.

- **Use a dash to indicate a strong afterthought that disrupts the sentence.**

  I know you're looking for — and I hope this helps — a list of qualified people.

Parentheses

Parentheses (often considered weak dashes) are like a sideshow; they're used to enclose one or more words in a sentence that aren't essential to the meaning of the sentence. Some examples of when to use parentheses follow:

- **Use parentheses around an expression that you want to de-emphasize.** A parenthetical expression is one that doesn't change the meaning of the sentence — that is, removing the expression doesn't alter the gist.

  This application (as unbiased researchers have established) is more powerful than what you're using.

- **Use parentheses around references to charts, pages, diagrams, authors, and so on.**

  Please read the section on fossils (pages 36–52).

- **Use parentheses to enclose numerals or letters that precede items in a series.**

  We are hoping to (a) get the draft completed by May 5, (b) get feedback by May 8, and (c) go to press on June 5.

When you enclose a sentence in parentheses, punctuate it as a sentence.

Brackets

Brackets aren't substitutes for parentheses. They have their own place in the world, as the following guidelines explain:

- **Use brackets to enclose words that you add to a direct quote.**

  He said, "The length of the study [from January to November] was entirely too long."

- **Use brackets as parentheses within parentheses.**

  Your order (including one dozen blue pens [which aren't available], three dozen green pens, and five dozen red pens) will ship on Monday, September 8.

Other Punctuation

Still to come are quotation marks, apostrophes, ellipses, hyphens, question marks, exclamation marks, periods, and slashes.

Quotation marks

Quotation marks are reserved for those occasions when you're citing something verbatim. If you paraphrase, don't use quotation marks.

> ***Quoting:*** Mr. Schultz said, "Please come to the meeting at 2:00."
>
> ***Paraphrasing:*** Mr. Schultz asked her to come to the 2:00 meeting.

- **Use quotation marks to enclose direct quotes.**

  "The high tech industry is vital to the economy," said the CEO.

- **Use quotation marks to enclose articles from magazines, songs, essays, short stories, one-act plays, sermons, paintings, lectures, and so on.**

  A recent issue of *Physics Today* magazine contained an article "Career Opportunities in Optics."

- **Use quotation marks to set off words or phrases introduced by expressions (such as the word, known as, was called, marked, entitled, and so on).** Another option is to use italics.

  Quotes: See the drawing marked "Wafer configure utilization."

  Italics: See the drawing marked *Wafer configure utilization.*

- **Use single quotation marks around a quotation within a quotation.**

  The consultant said, "You would do well to heed Mr. Smith's advice: 'Give the public what it wants, and you will be in business for a long time.'"

Ellipses

Ellipses show that words or names are omitted in a quotation. Place the ellipses where the omission occurs. They're formed by typing three periods with a space between each set of periods. When ellipses end a sentence, you don't need a final period.

> ***Omission at the beginning:*** " . . . The experiment consisted of printing 24 wafers at pre-determined screenprint settings."
>
> ***Omission at the end:*** "A Fractional Factorial 2^3 blocked by room temperature was used. The experiment consisted of printing 24 wafers at predetermined screenprint settings . . ."
>
> ***Omission somewhere in between:*** "The experiment consisted of printing 24 wafers . . . for shorts, opens, and solderbump height."

Apostrophes

Apostrophes aren't flying commas; they show possession or omissions.

Possession

Possession refers to ownership, authorship, brand, kind, or origin. These guidelines demonstrate how to use apostrophes to show possession:

- **Apostrophes are used most commonly with nouns to show possession.** In the following sentences, which host would you prefer?

  The presenter called the guests names when they arrived.

  The presenter called the guests' names when they arrived.
- **Form the possessive case of a singular noun by adding an apostrophe.**

  The idea is Jim's brainchild.
- **Form the possessive of a regular plural noun (one ending in *s*) by adding an apostrophe after the *s*.**

  The Murphys' lab is closed.
- **Form the possessive of an irregular plural noun (one not ending in *s*) by adding an apostrophe and *s*.**

  The salespeople's territories are being divided.
- **To show joint ownership, add the apostrophe and *s* after the last noun. To show single ownership, add the apostrophe and *s* to each noun.**

  Joint ownership: Jim and Pat's company is issuing an IPO.

  Individual ownership: Jim's and Pat's lockers are on separate floors.
- **In hyphenated words, put the apostrophe at the end of the possession.**

  He borrowed his brother-in-law's computer.
- **To make an abbreviation possessive, put an apostrophe and *s* after the period. If the abbreviation is plural, place an apostrophe after the *s*.**

  The Smith Co.'s testing starts next month.

  Two M.D.s' opinions are needed.
- **Express time and measurement in the possessive case.**

  We'll have an answer in one week's time.

It's becoming commonplace to write the names of companies and publications without apostrophes. When in doubt, check it out.

Omission

Use apostrophes to show that letters (as in contractions) or numbers (as in *the '90s*) are missing. Some companies frown on using contractions and prefer the more formal style. Defer to the style of the company.

- **Use an apostrophe to form the plural of a number, letter, or symbol, or word used as a separate word.**

  His last name has two f's.

  Sally doesn't always pronounce her r's at the end of a word.

  No if's, and's, or but's.

- **Use an apostrophe to show possession of initialisms or acronyms.** Some writers eliminate the apostrophe when there's little chance of misreading.

  I used MPR's suggestions.

It's wise not to use contractions for electronic documents. People with low-resolution screens may see a distorted apostrophe.

Hyphens

Don't confuse hyphens (-) with em dashes (—). They're different species. Hyphens function primarily as spelling devices.

- **Use a hyphen to join compound words that come before a noun.** Compound words are two or more words used as a unit to describe a noun.

  The company gave the researchers a ten-day extension.

  The company gave the researchers an extension of ten days.

- **Use a hyphen for compound numbers and written-out fractions.**

  One hundred fifty-two people attended the meeting.

  This is three-fourths the annual revenue.

- **Use a hyphen between a prefix that ends with a vowel and a word beginning with the same vowel. (When in doubt, check it out.)**

  It was said to be a pre-existing condition.

  The television station pre-empted my favorite program.

Question marks

Question marks serve as stop signs. Although you probably use them correctly, there are a few tricky situations. Hopefully, these guidelines can help demystify them:

- **Use a question mark after a short, direct question that follows a statement.**

  You saw the requisition, didn't you?

- **Use a question mark after each item in a series of questions within the same sentence.**

  Which of the IS candidates has the most experience? Mary? Joe? Jeff?

  Another option is the following: Mary, Joe, or Jeff?

  Don't capitalize the questions unless the beginning word should be capitalized, such as in the first of the above examples where there are names.

- **Use a question mark enclosed in parentheses to express doubt.**

  He said the results are due on April 8. (?)

Exclamation points

Exclamation points are reserved for words or thoughts that show strong feelings or emotions, as the following examples demonstrate:

Please try to do better!

That was an inspiring talk. Congratulations!

Periods

Periods are the stop signs of punctuation. They slow you down before you go on to the next thought.

- **Use a period after a statement, command, or request.**

  Thank you for getting us the results so promptly.

- **Use a period after words or phrases that logically substitute for a complete sentence.**

  No, not at all.

- **Use periods when writing abbreviations, acronyms, or initialisms.** A number of dictionaries, however, are citing many — for example, IBM, FDIC, and CPA — without periods. When in doubt, check it out. For more information about abbreviations, acronyms, or initialisms, see Appendix C.

Slashes

These critters go by a variety of names: slant lines, virgules, bars, or shilling lines. They separate or show an omission, such as in care of (c/o) or without (w/o).

- **Use slashes in *and/or* expressions.**

  The IS/training departments will present the training.

- **Use slashes in Internet addresses.**

  `http://www.hungryminds.com`

Appendix B

Grammar's Not Grueling

You may remember, when you were a kid, asking your mother, "Mom, can Pat and me go to the movies?" Your mother replied, "That's Pat and I" — and she didn't give you the money until you corrected your grammar. Although you didn't think so then, your mother was doing you a favor. Poor grammar didn't get you far with your mother, and it doesn't get you far in the business world.

Test Your Grammatical Skills

Take a look at the following sentences and see if you notice any errors. If you find all the mistakes, you're a grammar guru. Tear out this chapter of the book and share it with a deserving coworker. You find the answers at the end of the chapter.

1. A group of 75 computer specialists are waiting for the test results.
2. With most of the votes counted, the winner was thought to be her.
3. Everyone in the room, including the president and vice president, is being asked to do their share.
4. What was the name of the speaker we had yesterday?
5. Dr. Allen, who specializes in kinetics, would certainly be interested if he was here now.
6. The MVC Technology Company is celebrating their 50th anniversary.

I don't get into the nitty-gritty details of every part of speech because I don't want to bore you to tears. Rather, I touch upon the most troublesome areas in alphabetical order.

Adjectives

Adjectives answer at least one of the following questions: *What kind? Which? What color? How many?* or *What size?* They're words, phrases, or clauses that modify, describe, or limit the noun or pronoun they describe. You can use adjectives to transform an ordinary sentence into a tantalizing one. This can be a nice touch in a technical document that may be otherwise dry.

Forms of adjectives

Adjectives take different forms, depending on the noun or nouns they modify.

- **Use a positive adjective when you're not comparing anything.**

  It's *warm* in the lab.

- **Use a comparative adjective when you're comparing two things.**

  It's *warmer* in the lab today than it was yesterday.

- **Use a superlative adjective when you're comparing three things or more.**

  It's the *warmest* the lab has been all week.

Absolute adjectives

Some adjectives are absolute; they either *are* or *aren't*. For example, one thing can't be rounder than something else. Either it's round or it's not. Following are some adjectives that are considered absolute:

Complete	Correct	Dead	Empty
Genuine	Parallel	Perfect	Right
Round	Stationary	Unanimous	Wrong

Express the comparative and superlative forms of absolute adjectives by adding "more nearly" or "most nearly." For example, Jason's assumption was more nearly correct than Jim's.

Compound adjectives

In many cases, you use a hyphen to join together two adjectives so they form a single description. Use a hyphen only when the compound adjective comes before the noun, not after.

Before the noun: a part-time job; a two- or three-year experiment

After the noun: a job that's part time; an experiment of two or three years

Here are two exceptions:

1. Eliminate the hyphen when you generally think of the words as a unit; for example, post office address, life insurance, word processing, and the like.
2. Don't put a hyphen between adjectives if the first one ends in *-est* or *-ly;* for example, newly elected officer, freshest cut flowers, and so forth.

Articles

Use *the* to refer to a specific item and *a* or *an* to a non-specific item. Use *a* when a consonant sound follows the *a* (*a* vector, *a* method); use *an* when a vowel sound follows (*an* equation, *an* inductor).

Adverbs

Just as adjectives add pizzazz to nouns, adverbs spice up verbs. Adverbs modify verbs, adjectives, or other adverbs. They answer one or more of these questions: *How? When? Why? How much? Where?* or *To what degree?* Adverbs take different forms for the positive, comparative, and superlative, just as adjectives do.

Adjectives ending in *-ly* may also function as adverbs — depending on what they're modifying.

Adjective: The professor's handwriting is legible.

Adverb: The professor writes legibly.

Conjunctions

Conjunctions connect two or more words, phrases, or clauses that are equal in construction and importance. Common conjunctions are *and, or, for, so, but, nor,* and *yet.*

For information about punctuating sentences that contain conjunctions, see Appendix A.

Correlative conjunctions

You use some conjunctions in pairs to join the elements of a sentence. The most often-used pairs are the following:

Both/and	So/as	Not only/but also	Whether/or
Either/or	Neither/nor	As/as	Whether/or not

Subordinate conjunctions

Clauses starting with the following words and phrases always function as adverbs, adjectives, or nouns.

After	Although	As	As if	As long as	As though
Because	Before	Even if	Except	If	In order that
Provided	Since	Than	That	Though	Unless
Until	When	Where	While		

Double Negatives

If you ever said, "I don't want no liver," what you said is that you do want liver. Two negatives equal a positive. Never use two negative words to express one positive idea.

Correct: I don't have any solutions.

Incorrect: I don't have no solutions.

Nouns

Nouns — although critical to every sentence — are probably the least sexy part of speech. They don't create any emotion or add flair to your thoughts; they're merely *people, places,* or *things*. Proper nouns are specific and capitalized. Common nouns aren't specific or capitalized.

Proper nouns: New York City; New York University; Main Street

Common nouns: the city, the university, the street

Collective nouns are groups. When groups act as units (companies, councils, audiences, faculties, unions, teams, juries, committees, and so on), use a singular verb. When members of the group act independently, use a plural verb.

> ***Acting collectively:*** The team is going to test the equipment.
>
> ***Acting individually:*** The team are conducting separate tests.

Prepositions

Prepositions show the relationship between words and sentences. Here are some common prepositions:

Above	About	Across	After	Along
Among	Around	At	Before	Behind
Below	Beneath	Beside	Between	Beyond
By	Down	During	Except	For
From	In	Inside	Into	Like
Near	Of	Off	On	Since
To	Toward	Through	Under	Until
Up	Upon	With	Within	

Pronouns

Pronouns are words that substitute for nouns. Pronouns must agree with the nouns they replace in person, number, and gender.

- **Use *I, you, he, she, it, we,* or *they* when the pronoun is the subject of the sentence, or when it follows any form of the verb *to be: be, am, is, are, was, were, been, being, will be, had been*, and so on.**

  I will call her when the magazine arrives.

  Is Sam there? Yes, this is he.

- **Use *me, you, him, her, it, us,* or *them* when the pronoun is the object of either the verb or a preposition. These pronouns tell you "who" or "what."**

One trick to use when two pronouns come together is to break the thought into two sentences. It works every time!

Correct: Tom and I participated in the experiment.

(Tom participated in the experiment. I participated in the experiment.)

Incorrect: Tom and me participated in the experiment.

- **Use a possessive pronoun to indicate possession, kind, origin, brand, authorship, and so on. The possessive pronouns are *my, mine, your, yours, his, her, hers, its, our, ours, their,* and *theirs.***

  The decision is completely his.

  The coffee lost its taste. ("It's" is only used when you mean "it is.")

A common error occurs when referring to nouns such as in the word *company.* Because company is a singular noun, the pronoun that replaces the company name must be singular.

Correct: The company is having *its* annual meeting on August 15.

Incorrect: The company is having their annual meeting on August 15.

Singular pronouns

Certain pronouns are always singular and take singular verbs and pronouns, including *anybody, anyone, anything, each, either, everybody, everyone, everything, much, neither, nobody, nothing, one, somebody, someone,* and *something.*

Everyone, including Pete and Jane, has *his* and *her* opinions.

Neither Gary's nor Bob's proposals *is* acceptable.

Who and whom

Who-and-whom cowards can mumble these words and hope that listeners won't notice their indecision. Writers don't have that luxury. But it is possible to think of *who* and *whom* in easy terms!

When you can substitute *he, she,* or *they,* use *who.* And when you can substitute *her, him,* or *them,* use *whom.* (For the latter, I generally plug in an *-m* ending.)

The company needs a person *who* knows the new software. (*He* or *she* knows the new software.)

Are you the person to *whom* I spoke yesterday? (I spoke to *her* or *him.*)

Verbs

Verbs are the most important part of sentences because they express actions, conditions, or states of being. Verbs make statements about the subjects and can breathe life into dull text.

Gerunds

Gerunds are words or phrases whose roots are verbs ending with *-ing.* Gerunds start out as verbs, but act as nouns. When gerunds are preceded by nouns or pronouns, the nouns or pronouns take the possessive form.

> I don't like *your giving* me such short notice.
>
> *Ted's yelling* is quite irritating.

Dangling participles

If your participles dangle, it's nothing to be ashamed of. The condition's curable. *Dangling participles* are nothing more than verbs that don't clearly or logically refer to the nouns or pronouns they modify. Participles can dangle at the beginning or end of sentences. The following shows how to undangle participles:

> ***Correct:*** While James attended the meeting, the computer malfunctioned.
>
> ***Incorrect:*** While attending the meeting, the computer malfunctioned. (Who attended the meeting? That thought dangles.)

Were and was

Have you ever fantasized about being someone else? The English language provides a verb for those fantasies. "I wish I *were* . . ." The verb *were* is often used to express wishful thinking or an idea that's contrary to fact. *Was,* on the other hand, indicates a statement of fact.

> She acts as if she *were* president of the company. (Wishful thinking.)
>
> If Charles *was* at the conference, I didn't see him. (He may have been there.)

Was is the past tense of *is*. Why am I mentioning the obvious? Because people often mistakenly use *was* for the present tense when referring to something that's already happened.

> ***Correct:*** I thoroughly enjoyed reading the report — even though it *is* 950 pages long.
>
> ***Incorrect:*** I thoroughly enjoyed reading the report — even though it *was* 950 pages long.

Split infinitives

Present tenses of verbs that are preceded by *to* are called infinitives. Don't put modifiers between *to* and the verb, or you split infinitives and sometimes confuse and distract your readers.

> ***Correct:*** The instructor wants *to read* all the papers carefully.
>
> ***Incorrect:*** The instructor wants to carefully read all the papers.

For all you Star Trek fans, Gene Roddenberry's words *"To boldly go . . ."* are grammatically incorrect.

Subject and verb agreement

One of the most basic rules in grammar is that the subjects and verbs of sentences must agree. Both must be singular or both must be plural. Although most situations are pretty straightforward, the following sentences demonstrate situations that may be a little tricky:

- **Don't be fooled by interrupting phrases.**

  Correct: The software, despite the new installation manuals, still takes several days to install. (The subject is *software*.)

  Incorrect: The software, despite the new installation manuals, still take several days to install.

- ***A, many, an, each,*** **and *every* always take a singular verb.**

  Each and every computer *has* a modem.

  Many a man *is* denied this chance.

- ***None, some, any, all, most,*** **and *fractions* are either singular or plural, depending on what they modify.**

  Half the shipment was misplaced. (The subject, *shipment,* is singular.)

  Half the boxes were misplaced. (The subject, *boxes,* is plural.)

- **When referring to the name of a book, magazine, song, company, or article, use a singular verb even though the name may be plural.**

  Little Women is a great classic.

  Wanderman & Greenberg is a fine team of attorneys.

- **When referring to an amount, money, or distance, use a singular verb if the noun is thought of as a single unit.**

  I think *$900 is* a fair price.

  There *are 10 yards* of wire on the reel.

- **When *or* or *nor* is used to connect a singular and plural subject, the verb must agree in number with the person or item that is closest to the verb.**

  Neither Jim nor his *assistants were* available.

  Neither the assistants nor *Jim was* available.

Answers to Quickie Quiz

1. A group of 75 computer specialists *is* waiting for the test results. (*Group* is a singular subject and takes a singular verb.)
2. With most of the votes counted, the winner was thought to be *she*. (*She* was thought to be the winner. The nominative case is used when the pronoun is the subject of the sentence or when it follows any form of the verb *to be*.)
3. Everyone in the room, including the president and vice president, is being asked to do *his or her* share. (*Everyone* takes a singular verb, even if you throw in specific people. After all, the president and vice president are part of everyone. A better way of casting the sentence to avoid the clumsy *his/her* is to say: All the people in the room are being asked to do their share.)
4. What *is* the name of the speaker we had yesterday? (The speaker's name hasn't changed. Why use the past tense?)
5. Dr. Allen, who specializes in kinetics, would certainly be interested if he *were* here now. (Not a statement of fact. He isn't here now.)
6. The MVC Technology Company is celebrating *its* 50th anniversary. (The MVC Technology Company is a singular subject and takes a singular verb.)

Appendix C

Abbreviations and Metric Equivalents

This appendix shows how to write abbreviations. It also has tables of chemical elements, postal abbreviations, and metrics and U.S. equivalents.

Writing Abbreviations

Write out names or expressions the first time they appear. Thereafter, use abbreviations. If you're absolutely sure the reader will understand your reference, however, there's no need to write it out. When you do need to explain your reference, here's how to present it:

> *Write out random access memory (RAM) the first time you mention it in a document. Thereafter, you may use RAM because you already identified the reference for the reader.*

Be careful of doing this in online documents because you never know where the reader jumps in. You may use a pop-up window as described in Chapter 18.

If you deliver a paper document to a wide range of readers who may or may not understand your abbreviations, consider including a glossary of terms and abbreviations at the end of your document.

Treat the company or organization name as the company treats it. For example, if a company uses *Company*, don't write *Co.* Check the company Web site or letterhead for accuracy. If you can't check it out, write it out.

Acronyms and initialisms

What's the difference between acronyms and initialisms? Acronyms are formed by combining the first letter of several words and pronouncing the resulting string of letters as a single word. Initialisms are also formed by combining the first letter of several words, but they're pronounced as separate letters — as in "IBM."

Acronyms

- ISO (International Organization for Standardization)
- LASER (light amplification by stimulated emission of radiation)
- PERT (Performance Evaluation and Review Technique)

Initialisms

- OCR (optical character recognition)
- UFO (unidentified flying object)
- ROI (return on investment)

In business, industry, education, and government, acronyms and initialisms are often used by people who work together. That's fine as long as the readers easily understand your frame of reference, but it's quite possible that the reference may not be comprehensible to those outside your magical kingdom. Also, certain acronyms mean different things to different people. For example, following are a few associations that the acronym "ABA" may represent:

- American Banking Association
- American Bar Association
- American Booksellers Association
- American Bowling Association

Chemical elements

Table C-1 shows the chemical elements, their symbols, and their atomic numbers.

Table C-1 Chemical Elements

Name	*Symbol*	*Atomic No.*	*Name*	*Symbol*	*Atomic No.*
Actinium	**Ac**	89	Barium	**Ba**	56
Aluminum	**Al**	13	Berkelium	**Bk**	97
Americium	**Am**	95	Beryllium	**Be**	4
Antimony	**Sb**	51	Bismuth	**Bi**	83
Argon	**Ar**	18	Boron	**B**	5
Arsenic	**As**	33	Bromine	**Br**	35
Astatine	**At**	85	Cadmium	**Cd**	48

Name	*Symbol*	*Atomic No.*	*Name*	*Symbol*	*Atomic No.*
Calcium	**Ca**	20	Iron	**Fe**	26
Californium	**Cf**	98	Krypton	**Kr**	36
Carbon	**C**	6	Lanthanum	**La**	57
Cerium	**Ce**	58	Lawrencium	**Lr**	103
Cesium	**Cs**	55	Lead	**Pb**	82
Chlorine	**Cl**	17	Lithium	**Li**	3
Chromium	**Cr**	24	Lutetium	**Lu**	71
Cobalt	**Co**	27	Magnesium	**Mg**	12
Copper	**Cu**	29	Manganese	**Mn**	25
Curium	**Cm**	96	Mendelvium	**Md**	101
Dysprosium	**Dy**	66	Mercury	**Hg**	80
Einsteinium	**Es**	99	Molybdenum	**Mo**	42
Erbium	**Er**	68	Neodymium	**Nd**	60
Europium	**Eu**	63	Neon	**Ne**	10
Fermium	**Fm**	100	Neptunium	**Np**	93
Fluorine	**F**	9	Nickel	**Ni**	28
Francium	**Fr**	87	Niobium	**Nb**	41
Gadolinium	**Gd**	64	Nitrogen	**N**	7
Gallium	**Ga**	31	Nobelium	**No**	102
Germanium	**Ge**	32	Osmium	**Os**	76
Gold	**Au**	79	Oxygen	**O**	8
Hafnium	**Hf**	72	Palladium	**Pd**	46
Helium	**He**	2	Phosphorus	**P**	15
Holmium	**Ho**	67	Platinum	**Pt**	78
Hydrogen	**H**	1	Plutonium	**Pu**	94
Indium	**In**	49	Polonium	**Po**	84
Iodine	**I**	53	Potassium	**K**	19
Iridium	**Ir**	77	Praseodymium	**Pr**	59

(continued)

Table C-1 *(continued)*

Name	*Symbol*	*Atomic No.*	*Name*	*Symbol*	*Atomic No.*
Promethium	**Pm**	61	Terbium	**Tb**	65
Protactinium	**Pa**	91	Thallium	**Tl**	81
Radium	**Ra**	88	Thorium	**Th**	90
Radon	**Rn**	86	Thulium	**Tm**	69
Rhenium	**Re**	75	Tin	**Sn**	50
Rhodium	**Rh**	45	Titanium	**Ti**	22
Rubidium	**Rb**	37	Tungsten	**W**	74
Ruthenium	**Ru**	44	Unnilhexium	**Unh**	106
Samarium	**Sm**	62	Unnilpentium	**Unp**	105
Scandium	**Sc**	21	Unnilquadium	**Unq**	104
Selenium	**Se**	34	Unnilseptium	**Uns**	107
Silicon	**Si**	14	Uranium	**U**	92
Silver	**Ag**	47	Vanadium	**V**	23
Sodium	**Na**	11	Xenon	**Xe**	54
Strontium	**Sr**	38	Ytterbium	**Yb**	70
Sulfur	**S**	16	Yttrium	**Y**	39
Tantalum	**Ta**	73	Zinc	**Zn**	30
Technetium	**Tc**	43	Zirconium	**Zr**	40
Tellurium	**Te**	52			

Postal abbreviations

The United States Postal Service requests that you use the two-letter state abbreviation in all mailings. Don't use periods. Table C-2 shows the abbreviations for the United States and its territories.

Table C-2 United States and Territories

State	*Postal Code*	*State*	*Postal Code*
Alabama	AL	Montana	MT
Alaska	AK	Nebraska	NE
Arizona	AZ	Nevada	NV
Arkansas	AR	New Hampshire	NH
California	CA	New Jersey	NJ
Canal Zone	CZ	New Mexico	NM
Colorado	CO	New York	NY
Connecticut	CT	North Carolina	NC
Delaware	DE	North Dakota	ND
District of Columbia	DC	Ohio	OH
Florida	FL	Oklahoma	OK
Georgia	GA	Oregon	OR
Guam	GU	Pennsylvania	PA
Hawaii	HI	Puerto Rico	PR
Idaho	ID	Rhode Island	RI
Illinois	IL	South Carolina	SC
Indiana	IN	South Dakota	SD
Iowa	IA	Tennessee	TN
Kansas	KS	Texas	TX
Kentucky	KY	Utah	UT
Louisiana	LA	Vermont	VT
Maine	ME	Virginia	VA
Maryland	MD	Virgin Islands	VI
Massachusetts	MA	Washington	WA
Michigan	MI	West Virginia	WV
Minnesota	MN	Wisconsin	WI
Mississippi	MS	Wyoming	WY
Missouri	MO		

Metric and U.S. Equivalents

The United States is probably the only country in the universe that isn't using the metric system. Tables C-3 through C-7 show some of the popular metric measurements and their U.S. equivalents.

Table C-3 Linear Measures

U.S. System	*Metric System*
1 inch	25.4 millimeters (mm)
1 inch	2.54 centimeters (cm)
1 foot	304.8 millimeters (mm)
1 foot	30.48 centimeters (cm)
1 foot	0.3048 meter (m)
1 yard (36 in.; 3 ft.)	0.9144 meter (m)
1 rod (16.5 ft.; 5.5 yds.)	5.029 meter (m)
1 statute mile (5,280 ft.; 1760 yds.)	1,609.3 meters (m)
1 statute mile (5,280 ft.; 1760 yds.)	1.6093 kilometers (km)
Metric System	***U.S. System***
1 millimeter (mm)	0.03937 in.
1 centimeter (cm)	0.3937 in.
1 meter (m)	39.37 in.
1 meter (m)	3.2808 ft.
1 meter (m)	1.0936 yds.
1 kilometer (km)	3,280.8 ft.
1 kilometer (km)	1,093.6 yds.
1 kilometer (km)	0.62137 mi.

Table C-4 Liquid Measures

U.S. System	*Metric System*
1 fluid ounce (fl. oz.)	29.673 milliliters (ml)
1 pint (16 fl. oz.)	0.473 liter (l)
1 quart (2 pints; 32 fl. oz.)	9.4635 deciliters (dl)
1 quart (2 pints; 32 fl. oz.)	0.94635 liter (l)
1 gallon (4 quarts; 128 fl. oz.)	3.7854 liters (l)
Metric System	***U.S. System***
1 milliliter (ml)	0.033814 fl. oz.
1 deciliter (dl)	3.3814 fl. oz.
1 liter (l)	1.0567 qt.
1 liter (l)	0.26417 gal.

Table C-5 Area Measures

U.S. System	*Metric System*
1 square inch (0.007 sq. ft.)	6.452 square centimeters (cm^2)
1 square inch (0.007 sq. ft.)	645.16 square millimeters (mm^2)
1 square foot (144 sq. in.)	929.03 square centimeters (cm^2)
1 square foot (144 sq. in.)	0.092903 square meters (m^2)
1 square yard (9 sq. ft.)	0.83613 square meters (m^2)
1 square mile (640 acres)	2.59 square kilometers (km^2)
Metric System	***U.S. System***
1 square millimeter (mm^2)	0.00155 square inches (sq. in.)
1 square centimeter (cm^2)	0.155 square inches (sq. in.)
centiare	10.764 square feet (sq. ft.)
square kilometer (km^2)	0.38608 square miles (sq. mi.)

Table C-6 Capacity

U.S. System	*Metric System*
1 cubic inch (0.00058 cu. ft.)	16.387 cubic centimeters (cc; cm^3)
1 cubic inch (0.00058 cu. ft.)	0.016387 liters (l)
1 cubic foot (1,728 cu. in.)	0.028317 cubic meters (m^3)
1 cubic yard (27 cu. ft.)	0.76455 cubic meters (m^3)
1 cubic mile (cu. mi.)	4.16818 cubic kilometers (km^3)
Metric System	***U.S. System***
1 cubic centimeter (cc; cm^3)	0.061023 cubic inches (cu. in.)
1 cubic meter (m^3)	35.135 cubic feet (1.3079 cu. yd.)
1 cubic kilometer (km^3)	0.23390 cubic miles (cu. mi.)

Table C-7 Metric Conversions

Metric to U.S.	*U.S. to Metric*
Length	
millimeters × 0.04 = inches	inches × 25.4 = millimeters
centimeters × 0.39 = inches	inches × 2.54 = centimeters
meters × 3.28 = feet	feet × 3.04 = meters
meters × 1.09 = yards	yards × 0.91 = meters
kilometers × 0.6 = miles	miles × 1.6 = kilometers
Volume	
milliliters × 0.03 = fluid ounces	teaspoons × 5 = milliliters
milliliters × 0.06 = cubic inches	tablespoons × 15 = milliliters
liters × 2.1 = pints	fluid ounces × 30 = milliliters
liters × 1.06 = quarts	cups × 0.24 = liters
liters × 0.26 = gallons	pints × 0.47 = liters
cubic meters × 35.3 = cubic feet	quarts × 0.95 = liters
cubic meters × 1.3 = cubic yards	gallons × 3.8 = liters

Metric to U.S.	***U.S. to Metric***
Mass	
grams × 0.035 = ounces	ounces × 28 = grams
kilograms × 2.2 = pounds	pounds × 0.45 = kilograms
metric tons × 1.1 = short tons	short tons × 0.9 = metric tons
Area	
square centimeters × 0.16 = square inches	square inches × 6.5 = square centimeters
square meters × 1.2 = square yards	square yards × 0.8 = square meters
square kilometers × 0.4 = square miles	square miles × 2.6 = square kilometers
hectares (ha) × 2.5 = acres	acres × 0.4 = hectares
(The hectare is not an official SI unit, but it is permitted.)	

Appendix D

Technical Jabberwocky

Imagine little Alice trying to decipher a weird language when she was thrust into the strange world of Wonderland. *Jabberwocky, galumphing,* and *outgrabe* were as strange to Alice as *.dll* (pronounced dill), *GIGO*, and *newbie* may be to you. You, however, have a distinct advantage Alice didn't have. This glossary explains many of the terms you may come across.

24/7: Available 24 hours a day, 7 days a week. When you call a 24/7 hotline, however, you may hear the following message: "Your call is very important to us. Please stay on the line, and your call will be answered as soon as we finish servicing all of North America."

@: In an e-mail message, the "at" sign separates the user name from the service provider.

abstract: A "snapshot" of a long report or article to help the readers decide if they want to read the long text.

alpha test: People in your company test a product to get the kinks out before the product goes out for beta testing (which may be done by customers who are willing to be "test" subjects).

ASP (Application Service Provider): Manages and delivers applications to multiple entities from data centers across a wide area network.

asynchronous: A transmission in which each character or byte is synchronized by adding start and stop bits.

bandwidth: The amount of data a network connection can transfer at a given time. For example, 56K bps sends 56,000 bits of data per second. The greater the bandwidth, the faster you can transfer data. It's good to have a large bandwidth to transfer large files, graphics, or sound.

BBS (Bulletin Board System): An online area for posting messages. It's like the electronic version of a cork-board and tacks.

beta test: After a product passes the alpha test, you have it tested live. That may include testing by customers who use the product with the view towards purchasing it.

bisynchronous: A transmission in which a block or group of characters moves together with a single start and stop.

bit (Binary digIT): The smallest component of data — a 1 or 0 in the binary number system.

bookmark: An electronic feature that allows users to create a list of topics to revisit. It's much like using markers in books to tag pages that you want to revisit.

Bps (Bits Per Second): How fast a modem can send and receive data.

browser: (Also, *Web browser.*) Software that translates HTML files, converts them into Web pages, and displays them on your computer screen. Some browsers can access e-mail and newsgroups and play sound and video files.

buffer: Temporary electronic storage.

bug: A sticky wicket that prevents your software from operating properly.

byte: A group of eight bits that the computer processes as one unit.

CBT (Computer-Based Training): Training done on the computer that's generally delivered on CD-ROM.

chat room: An Internet location where you can "chat" with people around the world by typing messages and receiving responses in real time. Chat groups are often divided by special interests. They're a great way to meet people who share your personal and professional interests.

.com: Internet extension for commercial sites.

context-sensitive help: Help that's available in online documents with the click of a mouse. It provides a quick means of accessing information.

cookie: An Oreo, gingersnap, or small file code stored by your computer that's used as an ID tag to monitor the Web sites you visit.

COTS (Commercial Off-the-Shelf Software): Software that's ready to use when you buy it off the shelf. It requires little or no customizing.

crash: What happens when your computer stops running.

CRM (Customer Relationship Management): Software that gives sales, marketing, customer service, and help desk folks access to the same database so they can serve customers more efficiently.

decrypting: Decoding a message that's been coded (encrypted).

.dll: Filename extension for dynamic-link library files.

.doc: Filename extension for documents.

.dot: Filename extension for document templates.

domain name: (Also, *URL.*) A unique Web address.

download: To copy a file electronically to your computer. Think of it as loading something *down* from cyberspace. (The opposite is *uploading.*)

DSL (Designated Service Line): A high-speed telephone service for computer systems.

.edu: Internet extension for educational sites.

encrypting: Coding a message so that only the recipient can read it.

euphemism: A word or phrase that substitutes for one that may be offensive or blunt.

.exe: Filename extension for executable program files.

executive summary: Condenses a report into purpose, findings, and recommendations. Meant for executives who don't have the time or interest to read the long version of the report.

extranet: An extension of an intranet that gives select customers, suppliers, business partners, or other outsiders access to part(s) of a company's intranet. For example, a company may give certain suppliers access to pricing information.

FAQ (Frequently Asked Questions): A document that lists commonly asked questions and answers. FAQs may be posted on Internet newsgroups or online help.

filename extension: Used to distinguish between file types, such as .doc or .exe. (Many of the popular filename extensions are listed alphabetically in this glossary.)

firewall: A way to protect files and programs from unwanted access. It's somewhat like an airport customs officer who controls what can enter the country.

FTP (File Transfer Protocol): The Internet service that transfers files from one computer to another.

GIF (Graphics Interchange Format): Graphic files that are compressed to minimize the time it takes to transfer them over standard phone lines.

GIGO (Garbage In, Garbage Out): Your output is only as good as your input. In other words, if you key in garbage, you get rotten results.

.gov: Internet extension for government sites.

HTML (Hypertext Markup Language): The standard language for documents on the Web.

HTTP (Hypertext Transfer Protocol): A protocol to prepare Web pages for display when the user clicks on a hyperlink.

.ini: Filename extension for initial settings files.

Internet: A network in cyberspace that links computer networks all over the world by satellite and telephone. The Internet connects users with service networks such as the World Wide Web and e-mail.

intranet: A network of computers that only authorized users within a company can access. For example, people can post an invitation to a company-wide event on an intranet instead of sending e-mail attachments to everyone.

ISP (Internet Service Provider): A business that provides connectivity to e-mail and the Internet, such as AOL and AT&T.

JPEG (Joint Photographic Experts Group — pronounced jay-peg): A graphics file format for displaying graphics on the Web. It has a higher level of compression than GIF but results in a lower quality.

jump: (Also, *Hypertext Jump* or *Link.*) A cross-reference that lets users navigate from one topic to another. It shows up as a dotted line underneath the text.

keyboarding: The computer-age term for typing.

keyword: Describes a search topic. For example, if you have topics named *Coke* and *Pepsi*, the topic may not have the word *soda.* You can let users retrieve Coke and Pepsi when they type the keyword *soda.*

LAN (Local Area Network): A computer network that covers local distances and uses specialized computers to link smaller networks together.

link: An online document that "jumps" you from one piece of information to another.

modular chunk: A topic that stands alone because users will access only the information they need.

navigation aids: Used in a Web site to help users find information. Navigation aids include buttons, hyperlinks, and more.

.net: Internet extension for general sites.

newbie: Someone who's new to using the Internet.

online: Connected to the Internet.

online document: A document is an organized body of information with words and graphics. Online is simply the delivery medium — a computer, rather than paper.

.org: Internet extension for nonprofit sites.

PDAs (Personal Digital Assistants or Personal Device Applications): Hand-held, wireless computers such as palm units.

pop-up window: This window is triggered by a dotted line underneath the text. It overlays the parent window and displays additional information, such as a definition, topic overview, shortcut, or brief explanation.

RAM (Random Access Memory): A temporary storage system for creating, loading, manipulating data, and running programs.

ROM (Read Only Memory): Internal storage that holds instructions to the system.

RTFM (Read the "Friendly" Manual): If you're so inclined, you may substitute another word for *friendly.*

sans serif: Fonts without feet — no ascenders or decenders. A popular sans serif font is Arial.

serif: Fonts with feet — ascenders and decenders. A popular serif font is Times New Roman.

SME (Subject Matter Expert): A resource person who has an in-depth knowledge of the subject matter.

SSL (Secure Sockets Layer): Encryption that protects your e-mail messages from ogling eyes.

synchronous: A transmission in which there's a constant spacing of time between the serial bits of information.

.tif: Filename extension for default picture files, called TIFF.

.txt: Filename extension for text files used in Notepad.

typeface: Another word for font.

upload: To send a file from your computer. Think of it as loading something *up* into cyberspace. (The opposite is *downloading.*)

URL (Uniform Resource Locator): A string of information that makes up the address that gets you to a Web site. The most common URL starts with *http://*. Another is *ftp://*.

virus: This isn't a case of taking two aspirin and calling the doctor in the morning. A computer virus is a self-replicating code that's written by people with devious minds who want to damage or destroy computer systems. A virus can be downloaded from the Internet or an e-mail file, or transferred by an infected floppy disk. Software "vaccines," such as the Norton or McAfee virus checkers, give you an early warning so that you can avoid getting infected.

WAN (Wide Area Network): A computer network that covers a wide distance and uses specialized computers to link smaller networks together.

WBT (Web-Based Training): Training that is delivered on the Web. It's great for information that needs to be updated regularly because changing and posting are easy.

World Wide Web: (Also, *the Web* or *www*): Created in 1989 by Tim Berners-Lee, of Switzerland, when he attempted to organize all the information on the Internet.

WYSIWYG: (Pronounced *wizzy-wig.*) What you see (on-screen) is what you get (on the printout).

Index

• Numbers & Symbols •

• A •

• B •

• C •

• E •

• F •

• G •

• J •

• K •

• L •

• M •

• N •

• O •

• T •

• U •

• V •

• Z •

Notes